T0192850

Compact Textbooks in Mathematics

Compact Textbooks in Mathematics

This textbook series presents concise introductions to current topics in mathematics and mainly addresses advanced undergraduates and master students. The concept is to offer small books covering subject matter equivalent to 2- or 3-hour lectures or seminars which are also suitable for self-study. The books provide students and teachers with new perspectives and novel approaches. They feature examples and exercises to illustrate key concepts and applications of the theoretical contents. The series also includes textbooks specifically speaking to the needs of students from other disciplines such as physics, computer science, engineering, life sciences, finance.

- **compact:** small books presenting the relevant knowledge
- **learning made easy:** examples and exercises illustrate the application of the contents
- **useful for lecturers:** each title can serve as basis and guideline for a semester course/lecture/seminar of 2–3 hours per week.

More information about this series at http://www.springer.com/series/11225

Leonid Berlyand · Volodymyr Rybalko

Getting Acquainted with Homogenization and Multiscale

 Birkhäuser

Leonid Berlyand
Mathematics and Materials Research Institute
Pennsylvania State University
University Park, PA, USA

Volodymyr Rybalko
Mathematical Division
B Verkin Institute for Low
Temperature Physics and
Engineering of National Academy
of Sciences of Ukraine
Kharkiv, Ukraine

ISSN 2296-4568   ISSN 2296-455X   (electronic)
Compact Textbooks in Mathematics
ISBN 978-3-030-01776-7   ISBN 978-3-030-01777-4   (eBook)
https://doi.org/10.1007/978-3-030-01777-4

Library of Congress Control Number: 2018957824

Mathematics Subject Classification: 35-01, 35B27, 35B40, 35Q35, 35Q74, 74Q05, 74Q15, 76M50

This book is published under the imprint Birkhäuser, www.birkhauser-science.com by the
registered company Springer Nature Switzerland AG
The registered company address is: Gewerbestrasse 11, 6330 Cham, Switzerland

To our late mothers

Preface

The idea of this book came about when L. Berlyand taught several graduate courses on Multiscale and Homogenization at Penn State University. The classes were typically a mixture of first year mathematics and engineering graduate students as well as some advanced undergraduates. Most of these students were not necessarily thinking of writing their thesis on multiscale or homogenization, but rather were interested in learning about these subjects. This was no surprise as Google search on multiscale shows several million hits and continues to grow every year. Although the concept of scale plays a fundamental role in the natural sciences, engineering, and computational sciences, it is not typically introduced in mathematical textbooks. Homogenization theory was "born out of the computational battle with fine scales" in the modeling of heterogeneous media in engineering, physical and biological sciences. Typical examples include composite materials, porous media, fractured materials, and suspensions/colloids. All of these problems possess fine scale structure whose resolution results in computational challenges. Mathematical models of such problems lead to PDE and variational problems with small parameters (corresponding to fine scales), and analysis of such problems requires the development of novel asymptotic techniques.

It turned out that while there are many advanced mathematical textbooks on homogenization and multiscale (and thousands of research papers), the students in those classes wanted an introduction "from scratch" that did not require any significant experience in PDEs, calculus of variations, or functional analysis. One of the first textbooks on homogenization [93] was written for such an audience but since then several areas such as Γ-convergence for nonlinear problems, stochastic homogenization as well as continuum limits for discrete problems have seen rapid developments. At this point we recognized the challenge of designing a one-semester course for a quick introduction to the subject. That put us between a rock and a hard place of making the presentation accessible for beginners and providing an idea about the broad scope of research with an overview of basic homogenization techniques.

Our goal is to help a novice graduate student navigate through a quite mature field. We attempt to present mathematical ideas and techniques well known to experts while avoiding technicalities as much as possible. This is done in two ways. First, several basic techniques (two-scale asymptotic expansions, two-scale convergence, compensated compactness for both periodic and stochastic problems) are introduced using a case study conductivity problem. This provides the simplest PDE and variational setting from which one can move on to more technically involved problems, e.g., arising in fluid mechanics and elasticity theory. Second, we use 1D settings to introduce the reader to a complex world of nonlinear homogenization via Γ-convergence for problems in both continuum and discrete settings. Further these ideas and approaches are illustrated by examples that are of interest for material scientists and engineers, e.g., double porosity and plasticity problems. Moreover, the proofs are intended to be easy to follow: no steps are skipped, and rigorous arguments are supplemented by intuitive explanations and figures. Also, we choose to sacrifice generality for the sake of simplicity of presentation. Instead of the most general formulations of the main results, we present particular cases that still convey the main ideas but are simpler to follow, e.g., we consider scalar rather than vectorial settings. In short, we present known results in a new way suitable for a one-semester course for the specific audience of beginning graduate students in applied mathematics and engineering. The presentation is self-contained so that the only prerequisites are undergraduate courses in PDEs and analysis, and all other necessary mathematics is contained in the Preliminaries section. The book contains many exercises (often with hints), which are designed to help the reader master the techniques and better understand the presentation. While our primary focus is on introducing mathematical tools and techniques, modeling is also emphasized. We present several fundamental multiscale models arising in applications, e.g., double porosity and plasticity models, as well as basic stochastic models of heterogeneous media, e.g., random checkerboard and Poisson cloud.

We use material from well-known books on homogenization theory such as [15–17, 33, 37, 40, 62, 69, 82, 93, 97]. We focus our attention on several basic asymptotic approaches in homogenization, so that a number of important developments in homogenization theory are not presented in our book. We refer the reader to the books and monographs [4, 20, 34, 44, 50, 58, 72, 75, 87, 88, 98] for further reading on variational bounds, network approximation, optimization, meso-scale approximation, and multi-scale approaches for random and deterministic media. We also mention the field of numerical homogenization that builds on analytical homogenization. While it is hardly possible to review the literature in this fast growing field, we mention only a few subfields and selected references (a comprehensive list may be a book by itself!): multiscale finite element methods (MsFEM) [45, 59, 60, 73], heterogeneous multiscale methods (HMM) [9, 44, 48], adaptive multiscale methods [81], PDEs with rough coefficients [21, 84–86], localization [11, 68], and Bayesian numerical homogenization [83]. The researchers entering the field of numerical homogenization would benefit from analytical perspective in this book.

Finally we mention that this course was taught by L. Berlyand at summer schools in Kyoto and Haifa as well as in Beijing (Renmin University) and Shanghai (SJTU).

We now outline the contents of the 11 chapters of the book.

In Chapter 1, *Preliminaries*, we recall necessary notions and results from functional analysis, PDEs, calculus of variations, and probability theory. Special attention is paid to the notion of weak convergence which plays a key role in homogenization theory. We also review several results from elementary analysis which are typically not part of standard undergraduate courses (e.g., convex analysis, semicontinuity).

In Chapter 2, *What Is Homogenization and Multiscale? First Examples*, we introduce and discuss the concept of scales and provide numerous examples. We also present the two-scale conductivity problem which is used as a case study throughout the textbook and state the homogenization theorem in arbitrary dimension. We introduce the reader to rigorous analysis by proving the homogenization theorem for a 1D conductivity "baby" problem and provide numerical examples. We conclude this chapter with an interesting example of an inverse homogenization problem for thermo-elasticity in a 1D setting.

Chapter 3, *Brief History and Surprising Examples in Homogenization*, is devoted to the colorful history of homogenization that dates back to giants such as Poisson, Faraday, Maxwell, Rayleigh, and Einstein. Here we also mention the origins of the modern homogenization theory that started in the 1960s and is actively developing nowadays. Finally we present several examples of surprising predictions by homogenization theory.

In Chapter 4, *Formal Two-Scale Asymptotic Expansions and the Corrector Problem*, the fundamental idea of separation of fast and slow variables is used in the analysis of the multidimensional case study conductivity problem. We state a stronger version of the homogenization theorem that provides the rate of convergence of the homogenization approximation. This is done by introduction of a correcting term in the so-called corrector problem.

In Chapter 5, *Compensated Compactness and Oscillating Test Functions*, we describe a general concept of proving convergence in PDE problems based on the so-called Div-Curl Lemma. In the framework of homogenization theory, this approach led to what is known as the energy method. It provides much shorter arguments dealing with the weak formulation of the problem when compared to direct a priori estimates for two-scale expansions originally developed for homogenization problems.

In Chapter 6, *Two-Scale Convergence*, we present yet another approach to establishing the homogenization limit by synergy between the direct two-scale expansion and the energy method. Although it is restricted to periodic problems, it provides a simple framework for many practical problems where direct asymptotic expansions become prohibitively cumbersome. In particular we describe an interesting example of homogenization in a double porosity model.

In Chapter 7, *2D explicit example of effective conductivity via convex duality*, we present an elegant example of explicit calculation of effective conductivity for a 2D, two-phase problem. We use this example to introduce the reader to the beautiful idea of convex duality, which applies to many areas of mathematics.

In Chapter 8, *Introduction to Stochastic Homogenization*, we attempt to introduce the reader to a field which is currently very active. While there are a number of mathematical books on periodic homogenization, this is not the case for stochastic homogenization which, in part, is due to complexity of the subject that requires dual expertise in both PDEs/calculus of variations and probability. In this chapter we explain the distinct features of stochastic vs periodic homogenization and prove a classical theorem on existence of the homogenized limit for problems with stationary and ergodic random coefficients. Before proving the theorem we introduce the reader to stochastic models of heterogeneous media. This is done by providing detailed descriptions of basic examples of the random checkerboard and the Poisson cloud. These examples are easy on the intuitive level, but their rigorous mathematical understanding requires significant effort for someone with no experience in stochastic modeling. We conclude the chapter by a brief discussion of recent developments and provide a number of references for further reading.

In Chapter 9, *Γ-Convergence in Nonlinear Homogenization Problems*, we introduce a very powerful tool in the study of nonlinear variational problems. This is done in an abstract framework because Γ-convergence applies far beyond homogenization, e.g., in the theory of phase transitions. Next we demonstrate how Γ-convergence works in nonlinear homogenization problems. Here we use a 1D setting to explain ideas while minimizing the technicalities. In the subsequent Chapter 10, *An Example of a Nonlinear Problem: Homogenization of Plasticity and Limit Loads*, we present rigorous analysis of a practical problem in dimensions $n \geq 2$. Here we are not using directly the classical Γ-convergence approach but rather introduce techniques inspired by Γ-convergence. We also demonstrate the power of convex duality techniques that are further extended from the linear conductivity problem presented in Chapter 7.

In Chapter 11, *Continuum Limits for Discrete Problems with Fine Scales*, we use Γ-convergence techniques to prove the discrete to continuum limit in a 1D setting. We emphasize the "range of interaction parameter" which provides a key difference with the Γ-limit in continuum problems from Chapter 9. We explain in detail the effect of convexification in the Γ-limit which applies to both discrete and continuum settings and explain crucial differences between homogenization for convex and non-convex Lagrangians.

State College, PA, USA   Leonid Berlyand
Kharkiv, Ukraine   Volodymyr Rybalko
May 2018

Acknowledgments

We express our sincere gratitude to the Penn State graduate students and postdocs who provided invaluable help in the preparation of the manuscript: H. Chi, R. Creese, O. Iaroshenko, O. Misiats M.S. Mizuhara, M. Potomkin, and S. Ryan. Special thanks to our colleagues A. Kolpakov and A. Piatnitski for many useful discussions and suggestions. V. Rybalko is grateful to the Mathematics Department at Pennsylvania State University for hospitality during his visits, which provided opportunities to work on this book. We are also grateful to the National Science Foundation for support of various homogenization projects, which inspired the writing of this book.

Introduction

The book is based on a one-semester course on homogenization and multiscale aimed at beginning mathematics and engineering graduate students. This course was taught at Penn State University, at summer schools in Kyoto and Haifa, as well as in Beijing (Renmin University) and Shanghai (SJTU). The goal is to navigate such students through a quite mature field and provide an idea of its broad scope. An overview of basic homogenization techniques well known to experts is presented. Generality of results is sacrificed for the sake of simplicity of presentation, e.g., a 1D setting is used to introduce the reader to nonlinear homogenization. The proofs and definitions are supplemented with intuitive explanations and figures to make them easier to follow. The presentation is self-contained so that the only prerequisites are undergraduate courses in PDEs and analysis. The book contains many exercises (often with hints), which are designed to help the reader master the techniques and better understand the presentation. Examples from materials science and engineering illustrate application of general mathematical approaches. Several fundamental multiscale models are introduced and discussed in detail.

State College, PA, USA   Leonid Berlyand
Kharkiv, Ukraine   Volodymyr Rybalko
May 2018

Contents

Preliminaries

© Springer Nature Switzerland AG 2018
L. Berlyand, V. Rybalko, *Getting Acquainted with Homogenization and Multiscale*,
Compact Textbooks in Mathematics,
https://doi.org/10.1007/978-3-030-01777-4_1

In this chapter we present some classical definitions and results without proofs. For further reading of the material presented in ▶ Sections 1.1 through 1.5, see [25]. For introduction to PDEs beyond ▶ Section 1.5 we also recommend the book [49]. More details on Landau symbols, defined in ▶ Section 1.6, can be found in [39, 41]. For further information concerning convex analysis and semicontinuous functions from ▶ Section 1.7, see [47, 91]. Finally, as an additional reading to ▶ Section 1.8 on Probability Theory we recommend, e.g., [56, 95].

1.1 Banach and Hilbert Spaces

A complete, normed vector space is called a *Banach Space*. The dual space V^* of a Banach space V is the space of all continuous linear functionals f on V. That is, the pairing $\langle g, f \rangle$ is a linear function of $g \in V$ and $|\langle g, f \rangle| \leq C\|g\|_V$ for some C independent of g (the best constant C provides the norm of f). The Banach space V is called reflexive if $V^{**} = V$. A Banach space H with an inner product $(x, y)_H$ is called a *Hilbert Space*, and the norm is defined by $\|x\|_H := \sqrt{(x, x)_H}$.

1.2 Distributions and Distributional Derivative

1.2.1 Test Functions

The set of test functions, denoted by $\mathscr{D}(\Omega)$, is the set of infinitely differentiable functions with compact support (a function has compact support if it is equal to zero outside a compact set $K \subset \Omega$, where Ω is a bounded domain in $\mathbb{R}^N$). Convergence on $\mathscr{D}(\Omega)$ is introduced as follows: let $\varphi_i \in \mathscr{D}(\Omega)$, $i = 1, 2, ...$; $\varphi_0 \in \mathscr{D}(\Omega)$. Then the convergence $\varphi_i \to \varphi_0$ in $\mathscr{D}(\Omega)$ as $i \to \infty$ means that the supports of all functions belong to the same compact set from Ω, and φ_i and all its derivatives converge uniformly to φ_0 and its corresponding derivatives.

1.2.2 Distributions

Now, let T be a linear continuous functional determined on $\mathscr{D}(\Omega)$, i.e., a map that assigns to every $\varphi \in \mathscr{D}(\Omega)$ a number $\langle T, \varphi \rangle$, which is the value of the functional T on this function φ in $\mathscr{D}(\Omega)$. Essentially, $\langle \cdot, \cdot \rangle$ is the dual coupling. This map is linear with respect to φ and $\langle T, \varphi_i \rangle \to \langle T, \varphi_0 \rangle$ if $\varphi_i \to \varphi_0$ in $\mathscr{D}(\Omega)$ as $i \to \infty$. Such functionals are called distributions or generalized functions on Ω. The set of distributions is denoted by $\mathscr{D}'(\Omega)$.

Every locally integrable function $f(x)$ on Ω generates a distribution $\tilde{f} \in \mathscr{D}'(\Omega)$ defined as follows:

$$\langle \tilde{f}, \varphi \rangle = \int_\Omega f(x)\varphi(x)dx.$$

If such an f exists, then $\tilde{f}$ is referred to as a *regular distribution*. If $T \in \mathscr{D}'(\Omega)$, its distributional derivative $\frac{\partial T}{\partial x_i} \in \mathscr{D}'(\Omega)$ is determined by the equality

$$\left\langle \frac{\partial T}{\partial x_i}, \varphi \right\rangle = -\left\langle T, \frac{\partial \varphi}{\partial x_i} \right\rangle \tag{1.1}$$

for every $\varphi \in \mathscr{D}(\Omega)$. Formula (1.1) of the distributional derivative reminds us of the integration by parts identity. This is not a coincidence. Historically, the notion of the distributional derivative was inspired by this formula.

1.3   Functional Spaces

We consider a function $f(x)$ defined on the domain Ω such that $|f(x)|^p$ is Lebesgue integrable. This set of functions is denoted by $L^p(\Omega)$. When supplied with the norm

$$\|f\|_{L^p} = \left(\int_\Omega |f(x)|^p dx \right)^{\frac{1}{p}}$$

$L^p(\Omega)$ is a Banach space. If Ω is not bounded, we also use the spaces $L^p_{loc}(\Omega)$ consisting of functions $f \in L^p(K)$ for all compact subsets $K \subset \Omega$. Then convergence in $L^p_{loc}(\Omega)$ is understood as

$$f_n \to f \text{ in } L^p_{loc}(\Omega) \text{ if } f_n \to f \text{ in } L^p(K) \text{ for all compact subsets } K \subset \Omega. \tag{1.2}$$

The functional space $L^\infty(\Omega)$ is introduced as the set of measurable functions bounded almost everywhere. The norm $L^\infty(\Omega)$ is introduced as

$$\|f\|_{L^\infty} := \operatorname*{ess\,sup}_{x \in \Omega} |f(x)|$$

For $p = 2$, $L^2(\Omega)$ becomes a Hilbert space with the scalar product

$$(f, g)_{L^2} = \int_\Omega f(x)g(x)dx.$$

The Sobolev functional space $W^{m,p}(\Omega)$ is the set of distributions which (with all its distributional derivatives of order less than or equal to m) are generated by functions from $L^p(\Omega)$. This is a Banach space with the norm

$$\|f\|_{W^{m,p}} := \left(\int_\Omega \sum_{0 \leq m_1 + \ldots + m_N \leq m} \left| \frac{\partial^{m_1 + \ldots + m_N} f}{\partial x_1^{m_1} \ldots \partial x_N^{m_N}} \right|^p dx \right)^{\frac{1}{p}}. \tag{1.3}$$

For $p = 2$, $W^{m,p}(\Omega)$ is a Hilbert space, which is denoted by $H^m(\Omega)$, with scalar product

$$(f, g)_{H^m} = \int_\Omega \sum_{0 \leq m_1 + \ldots + m_N \leq m} \frac{\partial^{m_1 + \ldots + m_N} f}{\partial x_1^{m_1} \ldots \partial x_N^{m_N}} \cdot \frac{\partial^{m_1 + \ldots + m_N} g}{\partial x_1^{m_1} \ldots \partial x_N^{m_N}} dx. \tag{1.4}$$

For $m = 1$, the scalar product in $H^1(\Omega)$ is

$$(f, g)_{H^1} = \int_\Omega (f(x)g(x) + \nabla f(x)\nabla g(x)) \, dx, \tag{1.5}$$

where

$$\nabla = \left(\frac{\partial}{\partial x_1}, \ldots, \frac{\partial}{\partial x_n} \right). \tag{1.6}$$

The norm in $H^1(\Omega)$ is

$$\|f\|_{H^1} = \sqrt{\int_\Omega (f^2(x) + |\nabla f(x)|^2)dx}. \tag{1.7}$$

The functional space $H^m(\Omega)$ may also be introduced as the closure of the functional space $C^\infty(\Omega)$ in the norm (1.3) ($p = 2$). The closure of $\mathscr{D}(\Omega)$ (the space of $C^\infty(\Omega)$ functions with compact support) in the norm (1.3) or (1.7) is denoted by $W_0^{m,p}(\Omega)$ or $H_0^1(\Omega)$, respectively. In particular, this means that the sets $C^\infty(\Omega)$ and $\mathscr{D}(\Omega)$ are dense subsets of the respective functional spaces $H^m(\Omega)$ and $H_0^m(\Omega)$. Throughout this book we will frequently use the following space $H_{per(\Pi)}^1$ of periodic functions defined on the unit cube $\Pi = (0, 1)^n$. The space $H_{per(\Pi)}^1$ is defined as the closure of smooth Π-periodic functions, $C_{per}^\infty(\Pi)$, in the H^1 norm.

It is possible to introduce the space $H^s(\Omega)$ for real (not necessary integer or even positive) values of s. In the special case where $\Omega = \mathbb{R}^N$, the space $H^s(\mathbb{R}^N)$ can be introduced by using the Fourier transform. Here, $H^s(\mathbb{R}^N)$ is the set of all functions $f \in L^2(\mathbb{R}^N)$ such that their Fourier transforms

$$\hat{f}(\zeta) = (2\pi)^{-\frac{N}{2}} \int_{\mathbb{R}^N} f(x) e^{-i(x,\zeta)} dx, \qquad \zeta = (\zeta_1, ..., \zeta_N) \in \mathbb{R}^N \tag{1.8}$$

satisfy the condition $|(1 + |\zeta|^2)^{\frac{s}{2}} \hat{f}| \in L^2(\mathbb{R}^N)$. The norm of $f(x)$ in $H^s(\mathbb{R}^N)$ is introduced as follows:

$$\|f\|_{H^s(\mathbb{R}^N)} = \|(1 + |\zeta|^2)^{\frac{s}{2}} \hat{f}(\zeta)\|_{L^2(\mathbb{R}^N)}. \tag{1.9}$$

The idea of introducing of non-integer order derivatives is based on the well-known property of the Fourier transform of the derivative

$$i\zeta_n \hat{f}(\zeta) = (2\pi)^{-\frac{N}{2}} \int_{\mathbb{R}^N} \frac{\partial f}{\partial x_n}(x) e^{-i(x,\zeta)} dx, \qquad n = 1, ..., N \tag{1.10}$$

and Plancherel's theorem

$$\int_{\mathbb{R}^N} f^2(x) dx = \int_{\mathbb{R}^N} \hat{f}^2(\zeta) d\zeta. \tag{1.11}$$

From (1.10) and (1.11) we have

$$\int_{\mathbb{R}^N} |\nabla f|^2(x) dx = \int_{\mathbb{R}^n} |\zeta|^2 \hat{f}^2(\zeta) d\zeta. \tag{1.12}$$

For the second, third, and higher order derivatives:

$$(i)^2 \zeta_n \zeta_m \hat{f}(\zeta) = -\zeta_n \zeta_m \hat{f}(\zeta) = (2\pi)^{-\frac{N}{2}} \int_{\mathbb{R}^N} \frac{\partial^2 f}{\partial x_n \partial x_m}(x) e^{-i(x,\zeta)} dx, \tag{1.13}$$

and so on, and we conclude that the power m of the factor ζ in the expression $|\zeta|^m \hat{f}(\zeta)$ is the order of the derivatives of the original function $f(x)$.

It follows from (1.11) and (1.12) that the norm in $H^1(\mathbb{R}^N)$ can be defined in two equivalent ways:

$$\|f\|_{H^1}^2 = \int_{\mathbb{R}^N} \left(f^2(x) + |\nabla f(x)|^2 \right) dx = \int_{\mathbb{R}^n} (1 + |\zeta|^2) \hat{f}^2(\zeta) d\zeta. \tag{1.14}$$

The space $H^{-s}(Q)$, $s > 0$, is associated with the dual space of $H_0^s(Q)$.

ⓘ Remark 1.1 Note that

$$H_0^1(\Omega) \subset L^2(\Omega) \subset H^{-1}(\Omega).$$

If $f \in H^{-1}$, we note that the norm of f is defined

$$\|f\|_{H^{-1}(\Omega)} := \sup\{\langle f, u \rangle \mid u \in H_0^1(\Omega), \|u\|_{H_0^1(\Omega)} \leq 1\}.$$

The following theorem is often useful regarding characterization of the space $H^{-1}(\Omega)$:

Theorem 1.1

(i) *Assume $f \in H^{-1}(\Omega)$ with $\Omega \subset \mathbb{R}^n$. There exist functions $f^0, f^1, \dots, f^n \in L^2(\Omega)$ such that*

$$\langle f, v \rangle = \int_{\Omega} \left(f^0 v + \sum_{i=1}^{n} f^i v_{x_i} \right) dx, \quad \text{for all } v \in H_0^1(\Omega). \tag{1.15}$$

(ii) *Also*

$$\|f\|_{H^{-1}(\Omega)} = \inf \left\{ \left(\int_{\Omega} \sum_{i=0}^{n} |f^i|^2 dx \right)^{1/2} \mid f \text{ satisfies (1.15)} \right.$$

$$\left. \text{for } f^0, \dots, f^n \in L^2(\Omega) \right\}.$$

The following result about compact embeddings holds:

Theorem 1.2

Let $\Omega \subset \mathbb{R}^n$ be a bounded domain and $p \geq 1$. Then $W^{1,p}(\Omega)$ is compactly embedded in $L^p(\Omega)$, that is given an arbitrary bounded sequence $f_n \in W^{1,p}(\Omega)$, there exists a subsequence f_{n_k} which converges in $L^p(\Omega)$.

1.3.1 Trace Theorem

Consider a bounded set Ω with boundary $\partial\Omega$. Since functions in Sobolev spaces are only defined up to measure zero sets, restriction of Sobolev functions on Ω to $\partial\Omega$ must be understood using the *trace theorem*:

Theorem 1.3
For a domain Ω with a smooth boundary define the trace operator $T: C^\infty(\bar{\Omega}) \to C^\infty(\partial\Omega)$ by

$$Tu := u|_{\partial\Omega}.$$

If $\frac{1}{2} < s$, then T has a unique extension to a bounded linear operator

$$T: H^s(\Omega) \to H^{s-\frac{1}{2}}(\partial\Omega).$$

1.3.2 The Poincaré and the Friedrichs Inequalities

Let Ω be a bounded domain in $\mathbb{R}^n$. Then

Theorem 1.4 (The Friedrichs Inequality)
For every $u \in H_0^1(\Omega)$ it holds that

$$\int_\Omega u^2 dx \le C_f \int_\Omega |\nabla u|^2 dx, \tag{1.16}$$

with a constant C_f independent of u.

ⓘ **Remark 1.2** If $u = 0$ only on a part Γ of the boundary with $|\Gamma| = \int_\Gamma d\sigma > 0$, then bound (1.16) still holds with another constant $\tilde{C}_f$ depending on Γ.

If, additionally, Ω is connected and has a piecewise Lipschitz boundary, then

Theorem 1.5 (The Poincaré Inequality)
For every $u \in H^1(\Omega)$ such that $\int_\Omega u\,dx = 0$ the following holds:

$$\int_\Omega u^2 dx \le C_p \int_\Omega |\nabla u|^2 dx,$$

with a constant C_p independent of u.

1.4 Weak Convergence

In this section we recall the concept of weak convergence. Let V be a Banach space and V^* its dual, with a duality pairing $\langle \cdot, \cdot \rangle$. Consider a sequence $f_n \in V$ and $f_0 \in V$. We say that f_n *weakly converges* to f in V, written $f_n \rightharpoonup f_0$, if

$$\langle g, f_n \rangle \to \langle g, f_0 \rangle \text{ as } n \to \infty, \ \forall g \in V^*. \tag{1.17}$$

ⓘ Remark 1.3 If $\| f_n \|_V < C$, then for weak convergence it is sufficient to show (1.17) for all $g \in G \subset V^*$ such that G is dense in V^*.

ⓘ Lemma 1.1 *Let V be a Banach space and suppose that $f_n \in V$ converges to $f_0 \in V$ strongly. That is, $\| f_n - f \|_V \to 0$ as $n \to \infty$. Then $f_n \rightharpoonup f$ weakly in V. That is, strong convergence implies weak convergence.*

Proposition 1.1 *Let $\| f_n \|_V \leq C$ be a uniformly bounded sequence in a reflexive Banach space V. Then there exists a weakly convergent subsequence $\{ f_{n_k} \}_k$: $f_{n_k} \rightharpoonup f$.*

For our consideration the following observation is important. Let $\{f_n\}$, $\{g_n\}$ be functional sequences, e.g. $f_n, g_n \in L^2(0, 1)$, $f_n \rightharpoonup f$, $g \rightharpoonup g$ and $|f_n| < C$ (pointwise uniformly bounded). What can we say about the weak limit of their product $\lim_{n \to \infty} f_n g_n$ (if it exists)? The regular product rule does not necessarily hold (in contrast with strong convergence or pointwise limits).

Example 1.1

Suppose $v_n(x) = \sin(2\pi n x)$, then $v_n \rightharpoonup 0$ in $L^p(0, 1)$ for every $p > 1$. The graph of v_n for $n = 1$, $n = 2$, is depicted on the ◻ Figure 1.1. Observe that the oscillations become more dense over the domain as n increases. Due to the increase in the rate of oscillation as $n \to \infty$, we see that there is neither pointwise convergence nor strong convergence in $L^p(0, 1)$. For example, note in the case $p = 2$ that

$$\| \sin(2\pi n x) - \sin(2\pi m x) \|_{L^2}^2 = 1$$

for all $n, m \geq 1$. As such, the sequence $v_n = \sin(2\pi n x)$ is not a Cauchy sequence and thus cannot converge in norm. Nevertheless v_n *weakly* converges to zero, which is clear either by direct computation or by the Averaging Lemma in ▶ Section 2. ∎

◻ Fig. 1.1 Graph of the function $\sin(2\pi nx)$ for $n = 1$ (yellow) and $n = 2$ (green).

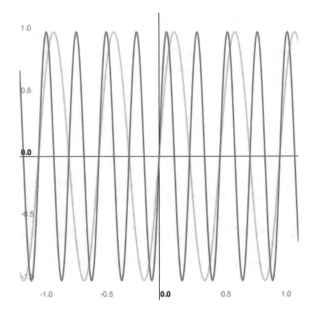

Exercise 1.1

Find the weak L^2-limit of $\sin \frac{x}{\varepsilon}$ and $\sin \frac{x}{\varepsilon} \cdot \sin \frac{x}{\varepsilon}$ as $\varepsilon \to 0$.   ∎

Let V be a Banach space and V^* be its dual space. A sequence $\{g_n\} \subset V^*$ converges in the weak* sense to $g_0 \in V^*$ if

$$\langle g_n, f \rangle \to \langle g_0, f \rangle \ \forall f \in V.$$

Exercise 1.2

Show that

- weak convergence implies weak* convergence, i.e. if $g_n \rightharpoonup g_0$ weakly in V^*, then $g_n \rightharpoonup g_0$ weak*;
- weak* convergence implies boundedness, i.e. $g_n \rightharpoonup g_0$ weak*, then $\limsup \|g_n\| < \infty$;
- If V is reflexive, then the notions of weak and weak* convergence coincide.

∎

For an example of reflexive Banach spaces, consider the space L^p. For p and q such that

$$\frac{1}{p} + \frac{1}{q} = 1, \ \ 1 < p < \infty$$

it is known that $(L^p)^* = L^q$. Since

$$(L^p)^{**} = (L^q)^* = L^p,$$

L^p is a reflexive Banach space. For $p = 1$ and $q = \infty$ we have $(L^1)^* = L^\infty$, however $(L^\infty)^* \neq L^1$ and one can construct a (bounded) linear function on L^∞, which cannot be represented as an integral with L^1 function. Both spaces L^1 and L^∞ are not reflexive.

1.5 Weak Solutions of Partial Differential Equations with Discontinuous Coefficients

Consider the following mixed boundary-value problem in a bounded domain $\Omega \subset \mathbb{R}^n$,

$$\operatorname{div} (A(x)\nabla u) = f(x) \text{ in } \Omega \tag{1.18}$$

$$(A(x)\nabla u(x) \cdot v(x)) = u_1(x) \text{ on } \Gamma \subseteq \partial\Omega, \tag{1.19}$$

$$u(x) = u_2(x) \text{ on } \partial\Omega \setminus \Gamma \tag{1.20}$$

with a symmetric tensor (matrix) $A(x) = (a_{ij}(x))_{i,j=1,\ldots n}$, $a_{ij}(x) = a_{ji}(x)$, whose entries $a_{ij}(x)$ are bounded

$$a_{ij} \in L^\infty(\Omega) \tag{1.21}$$

and satisfy the coercivity condition:

$$\exists \gamma > 0 \text{ s.t. } (A(x)\xi \cdot \xi) \geq \gamma|\xi|^2 \text{ for all } x \in \Omega \text{ and } \xi \in \mathbb{R}^n. \tag{1.22}$$

If the coefficients $a_{ij}(x)$, functions $f(x)$, $u_1(x)$, $u_2(x)$, surface $\partial\Omega$, and the boundary $\partial\Gamma$ of Γ on Ω are sufficiently smooth, the problem (1.18) has a classical solution $u \in C^2(\Omega) \cap C^1(\overline{\Omega})$. Multiplying the differential equation (1.18) by an arbitrary function $\varphi \in C^\infty(\overline{\Omega})$ such that $\varphi(x) = 0$ on $\partial\Omega \setminus \Gamma$ and integrating by parts, we obtain

$$\int_\Omega (A(x)\nabla u(x) \cdot \nabla\varphi(x)) \, dx = -\int_\Omega f(x)\varphi(x) dx + \int_\Gamma u_1(x)\varphi(x) d\sigma. \tag{1.23}$$

Both (1.20) and (1.23) are also well defined if $a_{ij} \in L^\infty(\Omega)$, $u_1 \in H^{-\frac{1}{2}}(\partial\Omega)$, $u_2 \in H^{\frac{1}{2}}(\partial\Omega)$, $u \in H^1(\Omega)$, $f \in H^{-1}(\Omega)$ and $\varphi \in H^1(\Omega)$. In this case, if (1.23) holds for all $\varphi \in H^1(\Omega)$ such that $\varphi = 0$ on $\partial\Omega \setminus \Gamma$, then u is called a weak solution of (1.18)–(1.20). In general this solution is not differentiable everywhere in Ω (in particular, for piecewise continuous coefficients $a_{ij}(x)$) but satisfies the equation (1.18) in the sense of distributions.

1.5.1 Variational Form of Boundary Value Problems

There is a direct link between solutions of (some) boundary value problems and variational problems. In particular, the problem (1.18)–(1.20) is related to the problem of minimization of the quadratic functional

$$I(u) = \frac{1}{2} \int_{\Omega} (A(x)\nabla u(x) \cdot \nabla u(x)) \, dx + \int_{\Omega} f(x)u(x)dx - \int_{\Gamma} u_1(x)u(x)d\sigma$$

(1.24)

among functions $u \in H^1(\Omega)$ such that $u(x) = u_2(x)$ on $\partial\Omega \setminus \Gamma$. Namely, if u is a minimizer, then taking test functions $u + \varepsilon\varphi$ (with $\phi = 0$ on $\partial\Omega \setminus \Gamma$) and equating to 0 the derivative of $I[u + \varepsilon v]$ in ε at $\varepsilon = 0$ (the necessary condition of minimum) one arrives at (1.23). Moreover every weak solution u of (1.18)–(1.20) is a minimizer of the functional $I[\cdot]$:

$$I[u+\varphi] = I[u] + \int_{\Omega} (A(x)\nabla u(x) \cdot \nabla\varphi(x)) \, dx + \int_{\Omega} f(x)\varphi(x)dx - \int_{\Gamma} u_1(x)\varphi(x)d\sigma$$

(1.25)

$$+ \int_{\Omega} (A(x)\nabla\varphi(x) \cdot \nabla\varphi(x)) \, dx = I[u] + \int_{\Omega} (A(x)\nabla\varphi(x) \cdot \nabla\varphi(x)) \, dx \geq I[u],$$

(1.26)

where $\varphi \in H^1(\Omega)$ is an arbitrary function vanishing on $\partial\Omega \setminus \Gamma$. In the particular case of the Poisson equation this equivalence between the PDE and variational problems is the well-known Dirichlet principle [49].

1.5.2 Lax-Milgram Theorem

Consider a real Hilbert space V with a norm $\|u\|_V$, and let V^* be a dual space with respect to some duality pairing $\langle \cdot, \cdot \rangle$. A scalar function $a(u, v)$ determined on $V \times V$ is called a bilinear form on V if it is linear with respect to the first and the second variables. A bilinear form $a(u, v)$ is called continuous if

$$|a(u, v)| \leq C\|u\|_V \|v\|_V \quad \forall u, v \in V.$$

(1.27)

A bilinear form $a(u, v)$ is called *coercive* if

$$\exists \gamma > 0 \text{ s.t. } |a(u, u)| \geq \|u\|_V^2 \quad \forall u \in V.$$

(1.28)

Theorem 1.6 (Lax-Milgram)
Let $a(u, v)$ be a continuous and coercive bilinear form, then for every $f \in V^*$ the problem

$$a(u, v) = \langle f, v \rangle \quad \forall v \in V,$$

(1.29)

has a unique solution $u = u(f)$.

This result applies to the problem (1.18)–(1.20) to establish existence and uniqueness of the solution in the case when the Dirichlet condition (1.20) is imposed on the surface Γ with $|\Gamma| = \int_\Gamma d\sigma > 0$. Introduce the bilinear form $a(u, v)$ by

$$a(u, v) = \int_\Omega (A(x)\nabla u(x) \cdot \nabla v(x))\, dx, \tag{1.30}$$

and fix a function $\tilde{u} \in H^1(\Omega)$ such that $\tilde{u} = u_2$ on $\partial\Omega \setminus \Gamma$. Then (1.23) rewrites in the form

$$a(u - \tilde{u}, v) = -\int_\Omega (A(x)\nabla\tilde{u}(x) \cdot \nabla v(x))\, dx - \int_\Omega f(x)v(x)dx + \int_\Gamma u_1(x)v(x)d\sigma. \tag{1.31}$$

Introducing the new unknown $w = u - \tilde{u}$ and the subspace V of functions from $H^1(\Omega)$ vanishing on $\partial\Omega \setminus \Gamma$ we can apply Theorem 1.6. Indeed, the right-hand side of (1.31) defines a bounded linear functional $\langle f, v \rangle$ on V while the bilinear form $a(w, v)$ is coercive, see Remark 1.2, and bounded.

In order to pass from weak formulation (1.31) to the classical formulation (1.18) one integrates by parts and uses the following fundamental lemma of calculus of variations, also known as the du Bois-Reymond Lemma.

ⓘ Lemma 1.2 (du Bois-Reymond) *Let $\Omega \subset \mathbb{R}^n$. If $f \in L^1_{loc}(\Omega)$ satisfies*

$$\int_\Omega f(x)v(x)dx = 0 \tag{1.32}$$

for all $v \in \mathscr{D}(\Omega)$, then $f \equiv 0$.

1.6 Landau Symbols O and o

The big "O" symbol together with its sibling little "o" are very useful notions in asymptotic techniques. Their definitions are briefly recalled below:

- Big O: $\phi(\varepsilon) = O(\psi(\varepsilon))$ if and only if there is a constant C such that $|\phi(\varepsilon)| < C|\psi(\varepsilon)|$ for sufficiently small ε. Example: $\sin(\varepsilon) = O(\varepsilon)$ and $e^\varepsilon = 1 + \varepsilon + O(\varepsilon^2)$ as $\varepsilon \to 0$.
- Big O of 1: $\phi(\varepsilon) = O(1)$ if and only if there is a constant C such that $|\phi(\varepsilon)| < C$ for sufficiently small ε. Example: $\sin(\varepsilon) = O(1)$ and $\log\varepsilon \neq O(1)$.
- Little o: $\phi(\varepsilon) = o(\psi(\varepsilon))$ if and only if $\phi(\varepsilon) = \gamma(\varepsilon)\psi(\varepsilon)$ and $\lim_{\varepsilon\to 0}\gamma(\varepsilon) = 0$. Example: $\varepsilon^3 = o(\varepsilon^2)$ and $\varepsilon\log\varepsilon = o(\varepsilon^\delta)$ for $\delta < 1$.
- Little o of 1: $\psi(\varepsilon) = o(1)$ if and only if $|\psi(\varepsilon)| \to 0$ as $\varepsilon \to 0$. Example.: $\varepsilon^\alpha = o(1)$ for all $\alpha > 0$.

1.7   Convex and Lower/Upper Semicontinuous Functions

Definition 1.1

A function $f : \mathbb{R} \to \mathbb{R}$ is called *convex* if

$$f(\lambda x + (1 - \lambda)y) \leq \lambda f(x) + (1 - \lambda)f(y) \tag{1.33}$$

for all $x, y \in \mathbb{R}$ and all $\lambda \in [0, 1]$.

Theorem 1.7 (Jensen's Inequality)
Assume $f : \mathbb{R} \to \mathbb{R}$ is convex and $u : (a, b) \to \mathbb{R}$ is integrable, then

$$f\left(\frac{1}{b-a} \int_a^b u(t)\, dt\right) \leq \frac{1}{b-a} \int_a^b f(u(t))\, dt. \tag{1.34}$$

Recall that convexity of a function that takes only finite values implies its continuity.

Definition 1.2

Let $X = (X, d)$ be a metric space. Consider a function $f : X \to \mathbb{R}$, and a point $x_0 \in X$. The function f is said to be *upper semi-continuous* at x_0 if

$$f(x_0) \geq \limsup_{x \to x_0} f(x). \tag{1.35}$$

Similarly f is said to be *lower semicontinuous* at x_0 if

$$f(x_0) \leq \liminf_{x \to x_0} f(x). \tag{1.36}$$

If the function is lower/upper semicontinuous at every point $x_0 \in \Omega \subset X$, then f is said to be lower/upper semicontinuous in Ω.

It is clear that if $f : X \to \mathbb{R}$ is both lower semi-continuous and upper semicontinuous at x_0, then f is continuous at x_0.

Theorem 1.8
Assume $K \subset \mathbb{R}$ is a compact set and $f : K \to \mathbb{R}$ is a lower semicontinuous function. Then there exists $x \in K$ such that $f(x) = \inf_{y \in K} f(y)$. That is, the infimum of f is attained on K.

Theorem 1.8 is a more general form of the well-known *Weierstrass Extreme Value Theorem*.

Proposition 1.2 (Lower Semicontinuity of Norms) *Let X be a Banach space and suppose that $\{x_n\} \subset X$ weakly converges to $x \in X$, then*

$$\liminf \|x_n\|_X \geq \|x\|_X. \tag{1.37}$$

Theorem 1.9 (Lower Semicontinuity of Integral Functionals)
Let $F : L^p(a, b) \to [0, +\infty]$ be given by

$$F(u) = \int_a^b f(u(x)) \, dx. \tag{1.38}$$

Then F is weakly lower semicontinuous in $L^p(a, b)$ if and only if f is lower semicontinuous and convex.

1.8 Elements of Probability Theory

Recall that given a set Ω, a collection of subsets $\mathscr{F}$ is called a σ-algebra if it satisfies the following:
1. (Contains universal set) $\Omega \in \mathscr{F}$.
2. (Closed under complements) If $F \in \mathscr{F}$, then $X \setminus F \in \mathscr{F}$.
3. (Closed under countable unions) If $F_1, F_2, \cdots \in \mathscr{F}$, then $\bigcup_{i=1}^{\infty} F_i \in \mathscr{F}$.

These subsets are called *measurable sets*. A function $f : \Omega \to \mathbb{R}$ is called *measurable* if the preimage of each of the sets $[a, \infty)$ is measurable. Given any arbitrary collection of subsets $\mathscr{C}$ of Ω the σ-algebra generated by $\mathscr{C}$ is the smallest σ-algebra containing $\mathscr{C}$ (see, e.g., [92]).

Definition 1.3

$(\Omega, \mathscr{F}, P)$ is a *probability space*, where Ω is a set referred to as the sample space containing all possible outcomes, $\mathscr{F}$ is a σ-algebra of measurable subsets, and P is a probability measure defined on $\mathscr{F}$, i.e. a measure such that $P(\Omega) = 1$.

Definition 1.4

$\xi : \Omega \to \mathbb{R}$, a measurable function from Ω to $\mathbb{R}$ is called a *random variable*.

Definition 1.5

Given a random variable ξ, function $P_\xi(\lambda) := P(\{\omega \in \Omega \mid \xi(\omega) \leq \lambda\})$ is called a *distribution function* of the random variable ξ.

Definition 1.6

The *expected value* or *probabilistic average* of a random variable $\xi(\omega)$ is given by

$$E(\xi) = \int_\Omega \xi(\omega) dP(\omega) = \int_{-\infty}^{\infty} \lambda \, dP_\xi(\lambda) \tag{1.39}$$

where the integrals in (1.39) are understood as Lebesgue-Stieltjes integrals (see [92]).

Definition 1.7

A random process is an indexed set of random variables ξ_τ, where τ is referred to as time, and typically $\tau \in \mathbb{R}$ (continuous time) or $\tau \in \mathbb{Z}$ (discrete time) . Finite dimensional joint distributions of a random process are defined by
$P(\xi_{\tau_1} \leq \lambda_1, \xi_{\tau_2} \leq \lambda_2, \ldots \xi_{\tau_N} \leq \lambda_N)$.

Definition 1.8

For fixed $\omega \in \Omega$ the function $f(t) = \xi_t(\omega)$ is called a *trajectory* or realization of the process.

Definition 1.9

A random field is a random process with n-dimensional time. In other words a random field $\xi_x(\omega)$ or $\xi(x, \omega)$ is a set of random variables, its realizations are random functions $\xi(x, \omega)$ of $x \in \mathbb{R}^n$.

Definition 1.10

A random field $\xi(x, \omega)$ is called *homogenous* or *stationary* if, for any finite set of points $x^1, ..., x^k \in \mathbb{R}^n$, and displacement $h \in \mathbb{R}^n$, the distribution function of the random vector

$$\left(\xi(x^1 + h, \omega), ..., \xi(x^k + h, \omega) \right),$$

is independent of h. In particular, random variables $\xi(x, \omega)$ have the same distribution functions for all $x \in \mathbb{R}^n$.

What Is Homogenization and Multiscale? First Examples

© Springer Nature Switzerland AG 2018
L. Berlyand, V. Rybalko, *Getting Acquainted with Homogenization and Multiscale*,
Compact Textbooks in Mathematics,
https://doi.org/10.1007/978-3-030-01777-4_2

2.1 Scales in Natural and Computational Sciences

We begin by introducing the concept of scale, and answering the question: why are scales important? Scales play a fundamental role in natural sciences, engineering, and computational sciences. The two main types of a scales are spacial scales (length scales) and temporal scales (time scales). A scale is not exactly length, but rather an order of magnitude. Examples of length scales are

$$\text{Atomic (micro) scale} \ : 10^{-10} m$$

$$\text{Planck (quantum) scale} \ : 10^{-35} m$$

In classical *homogenization* theory one typically deals with two scales; a microscale (fine scale) of the inhomogenous part of the composite material, and a macroscale or coarse scale of the size of the entire composite. The idea of homogenization, was born out of a computational battle with fine scales. The field has come a long way since then and has proven itself to be very useful. A practical goal of multiscale analysis is to replace equations with microscales (which are hard to solve numerically) by "averaged out" (coarse-grained or upscaled) macroscopic equations that are easier to solve.

Consider the following (classical) example of finding the conductivity of a composite material that consists of two different homogeneous materials (1 and 2). In this model, we consider material 2 to be the background medium (also called the matrix in the theory of composites), and the inclusions to be material 1, see ◘ Figure 2.1. When considering the conductivity of this whole composite, we must take the properties of both materials into consideration.

Assume that the background matrix of conductivity σ_2 has a large number of small periodically distributed inclusions of conductivity σ_1. Further assume that both media are isotropic so that σ_1, σ_2 are scalar quantities. Then, given the density g of the

Fig. 2.1 Two-scale composite.

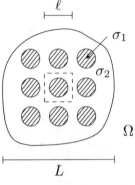

Fig. 2.2 Unit periodicity cell.

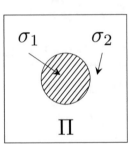

charge distributed in the composite material Ω, the electrostatic potential u satisfies the following boundary value problem:

$$-\operatorname{div}(\sigma(x)\nabla u) = g(x),\ x \in \Omega \tag{2.1}$$

$$u = U,\ x \in \partial\Omega, \tag{2.2}$$

where we assume that the boundary values U of the potential u are known, also the conductivity σ is given, $\sigma(x) = \sigma_2$ in the background matrix and $\sigma(x) = \sigma_1$ in inclusions. Note that if $\sigma_1 = \sigma_2$ then (2.1) becomes the Poisson equation $-\sigma \Delta u = g(x)$ in Ω, while (2.1) is a divergence form elliptic PDE of more general form.

The two scales in this example are the macroscale L (the sample size) and the microscale ℓ (the period of the inclusions). Since the number of inclusions is large $\ell \ll L$, and thus we introduce the dimensionless small parameter

$$\varepsilon = \frac{\ell}{L} \ll 1.$$

■ Figure 2.2 depicts the blown up periodicity cell in rescaled coordinated $y = x/\varepsilon$, for simplicity hereafter the periodicity cell Π is assumed to be the unit cube $\Pi = (0, 1)^n$, hereafter called the *unit periodicity cell* or simply the *periodicity cell*.

19

2

2.2 · Two-scale Case Study Conductivity Problem. Homogenization…

When parameter ε is small, one still can use numerical methods to find an approximate solution of the problem. However, employing standard techniques such as finite element and finite difference methods require the use of a very fine mesh, much finer than ε. Thus, in many real-life problems such as this one, these techniques become very expensive, to the point where their use is prohibited. The idea then is to use an asymptotic method and study behavior of solutions as ε tends to zero.

2.2 Two-scale Case Study Conductivity Problem. Homogenization Theorem and Averaging Lemma

Consider this question in a context of more general divergence form elliptic PDEs which describe anisotropic media (unlike problem (2.1)–(2.2)). Assume we are given a matrix

$$\sigma = \{\sigma_{ij}(y)\}_{i,j=1,\dots,n}$$

with entries $\sigma_{ij} \in L^\infty(\Omega)$ which possess the symmetry $\sigma_{ij} = \sigma_{ji}$, satisfy the ellipticity condition:

$$\exists \alpha > 0 \text{ such that } \alpha |\xi|^2 \le \sum_{ij} \sigma_{ij}(y)\xi_i\xi_j, \ \forall \xi \in \mathbb{R}^n, \tag{2.3}$$

and are bounded

$$\sum_{ij} \sigma_{ij}\xi_i\xi_j \le \beta |\xi|^2, \ \forall \xi \in \mathbb{R}^n. \tag{2.4}$$

Next consider the *case study* two-scale conductivity problem

$$-\operatorname{div}(\sigma(x/\varepsilon)\nabla u_\varepsilon) = g(x), \ x \in \Omega \tag{2.5}$$

$$u_\varepsilon = 0, \ x \in \partial\Omega. \tag{2.6}$$

Here we have a more general elliptic PDE with right hand side $g(x)$. For simplicity we prescribe zero Dirichlet data on the boundary (the boundary data are not important for the theory presented below). As before we consider periodic $\sigma(y)$, $\sigma \in L^\infty_{per}(\Pi)$, i.e. $\sigma \in L^\infty(\mathbb{R}^n)$ and $\forall i \in [1..n]$, $\sigma(y + \mathbf{e}_i) = \sigma(y)$ a.e. in $\mathbb{R}^n$, where $\{\mathbf{e}_i\}$ is the canonical basis in $\mathbb{R}^n$.

Since we do not assume that σ smooth (in the above example it is piecewise constant and discontinuous) the solution u_ε of (2.5)–(2.6) has to be understood in the weak (variational) sense even for smooth right hand sides $g(x)$. That is $u_\varepsilon \in H^1_0(\Omega)$ must satisfy

$$\sum_{i,j=1}^n \int_\Omega \sigma_{ij}(x/\varepsilon)\frac{\partial u_\varepsilon}{\partial x_j}\frac{\partial \phi}{\partial x_i}dx = \int_\Omega g(x)\phi(x)dx, \ \forall \phi \in H^1_0(\Omega). \tag{2.7}$$

This is a well-posed problem for every given $g \in L^2(\Omega)$ (or $g \in H^{-1}(\Omega)$) by the Lax-Milgram theorem (see preliminaries). Moreover, its unique solution u_ε satisfies the uniform bound

$$\|u_\varepsilon\|_{H^1(\Omega)} \leq C\|g\|_{L^2(\Omega)}. \tag{2.8}$$

Indeed, take $\phi = u_\varepsilon$, then

$$\alpha\|\nabla u_\varepsilon\|^2_{L^2(\Omega)} \overset{\text{by (2.3)}}{\leq} \int_\Omega \sigma_{ij}(x/\varepsilon)\frac{\partial u_\varepsilon}{\partial x_i}\frac{\partial u_\varepsilon}{\partial x_j}dx = \int_\Omega gu_\varepsilon$$

$$\leq \|g\|_{L^2(\Omega)}\|u_\varepsilon\|_{L^2(\Omega)} \quad \text{(by the Cauchy-Schwartz inequality)}$$

$$\leq C\|g\|_{L^2(\Omega)}\|\nabla u_\varepsilon\|_{L^2(\Omega)}. \quad \text{(by the Poincaré inequality)}$$

Now, divide both sides by $\alpha\|\nabla u_\varepsilon\|_{L^2(\Omega)}$ to obtain that $\|\nabla u_\varepsilon\|_{L^2(\Omega)} \leq C_1\|g\|_{L^2(\Omega)}$, and apply the Poincaré inequality once again to get (2.8).

Thanks to the a priori estimate (2.8) the limit of u_ε as $\varepsilon \to 0$ exists, at least up to subsequence. Indeed, since norms $\|u_\varepsilon\|_{H^1(\Omega)}$ are bounded, one can extract a subsequence weakly converging in $H^1_0(\Omega)$ to some limit $u_0(x)$. By the compactness of the embedding of $H^1_0(\Omega)$ into $L^2(\Omega)$ (see Theorem 1.2) this subsequence converges strongly in $L^2(\Omega)$. Note that $u_\varepsilon(x)$ solves a two-scale problem with fine scale of the order ε and coarse scale of the order 1. It turns out that every limit $u_0(x)$ solves a problem similar to (2.5)–(2.6) but with coefficients that no longer have fine scale:

$$-\text{div}\,[\hat{\sigma}\nabla u_0] = g(x), \qquad x \in \Omega \tag{2.9}$$

$$u_0 = 0, \qquad x \in \partial\Omega. \tag{2.10}$$

Since (2.9)–(2.10) has a unique solution, the entire sequence u_ε converges to u_0 in $L^2(\Omega)$. Also in the case of periodic coefficients $\sigma_{ij}(x/\varepsilon)$, elements $\hat{\sigma}_{ij}$ of the matrix $\hat{\sigma}$ are constants. The problem (2.9)–(2.10) is called the *homogenized* problem for the two-scale problem (2.5)–(2.6) and $\hat{\sigma}$ is called the *homogenized* or *effective* conductivity tensor. Since both names are widely used in literature we will also use both of them throughout the book.

The ideas of mathematical homogenization come from physics (which often happens in mathematics). We consider the example of the equivalent homogeneity principle from mechanics of fluids or solids [36]. It states that one can ignore the fact that fluids and solids are discrete systems made of atoms and suggests that such systems can be described by continuum PDE models instead of discrete models of interacting atoms (molecules), illustrated by ◻ Figure 2.3.

◻ **Fig. 2.3** Black dots on the left correspond to atoms in a lattice. The gray medium on the right is a continuum system.

2.2 · Two-scale Case Study Conductivity Problem. Homogenization...

21

2

□ **Fig. 2.4** Two-phase
composite is depicted on the left
and the homogenized medium
is depicted on the right.

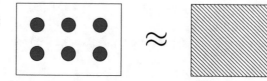

Recent development in homogenization theory known as the *discrete to continuum limit* provides a rigorous justification of this hypothesis that includes its limits of validity. It amounts to investigation of the discrete models in the limit when lattice spacing $\varepsilon \to 0$ (see ► Section 11). As an example, from a coupled large system of ODEs that models a discrete mass-spring system one can obtain a macroscopic PDE of elasticity. This PDE can contain more information that a simple phenomenological model based on the empirical Hooke's law.

To further explain, we look back to example (2.1)–(2.2). Recall from physics the notion of *homogeneous medium*. A medium is called *homogeneous* if its properties are identical at all points of the medium. In our example if $\sigma_1 = \sigma_2$, then the medium is homogeneous. If we have $\sigma(x) \neq$ constant, then we call the medium *heterogeneous*.

Example (2.1)–(2.2) is a two-phase composite case study with a periodic structure. It is equivalent to a homogeneous system by the equivalent homogeneity principle [36]. Mathematical homogenization theory makes this equivalence precise in the following sense: it establishes existence of the homogenized problem and convergence of solutions that justifies the equivalent homogeneity principle (□ Figure 2.4).

We now state a basic theorem that illustrates the homogenization procedure for the problem (2.5)–(2.6). This theorem is proved in ► Section 5 (and an alternative proof is given in ► Section 6).

Theorem 2.1

Let $u_\varepsilon(x)$ be the solution of the problem (2.5)–(2.6) in a bounded domain Ω. Then,

$$u_\varepsilon \rightharpoonup u_0 \text{ weakly in } H_0^1(\Omega) \tag{2.11}$$

$$\sigma_\varepsilon \nabla u_\varepsilon \rightharpoonup \hat{\sigma} \nabla u_0, \text{ weakly in } L^2(\Omega) \tag{2.12}$$

where $u_0(x)$ is the unique solution of the homogenized problem (2.9)–(2.10). The effective conductivity tensor $\hat{\sigma}$ is defined by

$$\hat{\sigma}_{ij} = \frac{1}{|\Pi|} \int_\Pi (\sigma(y)(\nabla \chi_i + e_i)) \cdot (\nabla \chi_j + e_j) dy. \tag{2.13}$$

(continued)

Theorem 2.1 (continued)
Where χ_i are the solutions of the so-called "cell problems":

$$-\text{div}\,[\sigma(y)[\nabla\chi_i + e_i]] = 0, \quad y \in \Pi \tag{2.14}$$

χ *and its first derivatives are Π-periodic.* $\qquad\qquad\qquad$ (2.15)

Here $1 \leq i, j \leq n$.

ⓘ **Remark 2.1** Note that from (2.11) we have $\nabla u_\varepsilon \rightharpoonup \nabla u_0$ in $L^2(\Omega)$. Further, it will be established in Lemma 2.1 that $\sigma_\varepsilon \rightharpoonup \langle\sigma\rangle$ in $L^2(\Omega)$. However, this is far not sufficient to deduce (2.12). Actually we will see that $\hat{\sigma} \neq \langle\sigma\rangle$.

Theorem 2.1 claims that $u_\varepsilon(x) \rightharpoonup u_0(x)$ weakly in $H^1(\Omega)$. Due to the Sobolev embedding theorems, weak H^1 convergence implies strong L^2 convergence. In other words,

$$\|u_\varepsilon(x) - u_0(x)\|_{L^2(\Omega)} \to 0 \text{ as } \varepsilon \to 0 \tag{2.16}$$

as well as weak L^2 convergence of $\nabla u_\varepsilon \rightharpoonup \nabla u_0$. However, most practical quantities (current, stress, etc.) involve derivatives of u_ε so we ideally wish to have strong L^2 convergence of $\nabla u_\varepsilon \to \nabla u_0$. However, typically it is not true that

$$\|u_\varepsilon - u_0\|_{H^1(\Omega)} \to 0. \tag{2.17}$$

Exercise 2.1
Assuming that Theorem 2.1 is proved, show that (2.17) does not hold if $\hat{\sigma} \neq \langle\sigma\rangle$. ∎

Let us return to the original discussion of scales, and how homogenization was born out of the computational battle with fine scales. A direct numerical solution of (2.5)–(2.6) is very expensive for small ε due to the ε scale oscillations. In this case, the computational mesh (or grid) must be very fine to resolve the problem with an acceptable accuracy. However, equations (2.9)–(2.10) and (2.14)–(2.15) do not depend on ε. Thus, the homogenized problem has only one macroscopic scale of the order 1, which is not expensive to resolve.

Note that solving the cell problem (2.14)–(2.15) is a much easier computational task than resolving (2.5)–(2.6) for the following reasons:

(i)   It only has a coarse scale (small parameter ε is no longer present).
(ii) Its solution does not depend upon the right hand side and it satisfies a standard, periodic boundary condition.
(iii) The cell problem does not depend upon the domain Ω since it is defined on the unit periodicity cell Π.

23

2

2.2 · Two-scale Case Study Conductivity Problem. Homogenization…

Thus we see that the cell problem encodes properties of microstructure. Once the cell problem is solved and $\hat{\sigma}$ is found, one can perform computation on the mesoscale h for any right hand side $g(x)$ and any domain Ω, and these computations will not require resolving the fine scale ε.

The solutions of the cell problem have *"coordinate-like behavior"* which can be illustrated as follows. Introduce $u_i(y) := \chi_i + y_i$, $y \in \Pi$. Then

$$-\text{div}\,[\sigma(y)\nabla u_i] = 0, \quad y \in \Pi$$

with $u_i - y_i$ being Π-periodic (with first derivatives).

The functions u_i "mimic" the coordinate functions y_i in the following sense: consider the simplest case $\sigma_1 = \sigma_2$. Then u_i defined by $u_i(y) := \chi_i + y_i$ is actually $u_i = y_i$ since $\chi_i = 0$. For the general case, when $\sigma_1 \neq \sigma_2$ the function u_i "tries to" minimize the deviation from the coordinate function y_i in the following sense. It deviates from y_i by the function χ_i which minimizes the energy functional

$$\int_\Pi \sigma(y)|\nabla(\chi_i + y_i)|^2 dy \tag{2.18}$$

in the class $H^1_{per}(\Pi)$ (cf. variational problems in Subsection 1.5.1).

The following exercise helps to develop intuitive understanding of the homogenized conductivity tensor.

Exercise 2.2

Consider a cubic cell with a spherical inclusion. Assume in addition that $\sigma(y)$ satisfies cubic symmetry: $\sigma(Ty) = \sigma(y)$ (where T is any symmetry of the cube; i.e., any linear transformation which maps the cube into itself). Using the definition of $\hat{\sigma}_{ij}$ show that

$$\sigma_{ij} = \hat{\sigma}\,\delta_{ij},$$

i.e., that the homogenized conductivity tensor is a scalar. ∎

A characteristic mathematical feature of our example (2.5)–(2.6) is the presence of *rapidly oscillating coefficients*. Note that rapidly oscillating functions lack pointwise convergence and thus require a notion of weak convergence (see Preliminaries Subsection 1.4). Using weak convergence we introduce a first tool to understand the limiting behavior of such rapidly oscillating functions.

ⓘ **Lemma 2.1 (Averaging Lemma)** *Let Ω be a bounded open set in $\mathbb{R}^n$, and let $g \in L^2_{loc}(\mathbb{R}^n)$ be a Π-periodic function, then*

$$g\left(\frac{x}{\varepsilon}\right) \rightharpoonup \langle g \rangle \quad \text{weakly in } L^2(\Omega) \text{ as } \varepsilon \to 0, \tag{2.19}$$

where $\langle g \rangle = \dfrac{1}{|\Pi|} \displaystyle\int_\Pi g(x)\,dx.$ $\tag{2.20}$

Proof

The set of smooth compactly supported functions $C_0^\infty(\Omega)$ is dense in $L^2(\Omega)$. Thus, it is sufficient to prove (2.19) for $\theta \in C_0^\infty(\Omega)$.

Next, observe that functions $g\left(\frac{x}{\varepsilon}\right)$ are uniformly bounded in $L^2(\Omega)$:

$$\int_\Omega \left|g\left(\frac{x}{\varepsilon}\right)\right|^2 dx = \sum_i \int_{\Pi_i^\varepsilon \cap \Omega} \left|g\left(\frac{x}{\varepsilon}\right)\right|^2 dx \le \frac{C_\Omega}{\varepsilon^n} \cdot \int_\Pi |g(y)|^2 \varepsilon^n \, dy \le C_\Omega \|g\|_{L^2(\Pi)}^2,$$

(2.21)

where Π_i^ε are (small) cells obtained by translating the scaled reference cell $\varepsilon\Pi$ to points $x_i^\varepsilon \in \varepsilon\mathbb{Z}^n \cap \Omega$, i.e. $\Pi_i^\varepsilon = x_i^\varepsilon + \varepsilon\Pi$. We must show that

$$\int_\Omega g\left(\frac{x}{\varepsilon}\right)\theta(x)dx \to \langle g\rangle \int_\Omega \theta(x)dx.$$

(2.22)

Define the function θ_ε to be piecewise constant interpolation of θ by

$$\theta_\varepsilon(x) = \theta(x_i^\varepsilon), \quad \text{when } x \in \Pi_i^\varepsilon.$$

(2.23)

Next, we use the Cauchy-Schwartz inequality to write

$$\int_\Omega \left|g\left(\frac{x}{\varepsilon}\right)(\theta(x) - \theta_\varepsilon(x))\right| dx \le \|g(x/\varepsilon)\|_{L^2(\Omega)}\|\theta - \theta_\varepsilon\|_{L^2(\Omega)}$$

$$\overset{\text{by (2.21)}}{\le} C\|g\|_{L^2(\Pi)}\|\theta - \theta_\varepsilon\|_{L^2(\Omega)}.$$

(2.24)

Note that $\|\theta(x) - \theta_\varepsilon(x)\|_{L^2(\Omega)} \to 0$ as $\varepsilon \to 0$. Since θ has compact support in Ω, for sufficiently small ε we have $\theta = \theta_\varepsilon = 0$ in cells Π_i intersecting with the boundary $\partial\Omega$, therefore

$$\int_\Omega g\left(\frac{x}{\varepsilon}\right)\theta_\varepsilon(x)\, dx = \sum_i \int_{\Pi_i^\varepsilon} g\left(\frac{x}{\varepsilon}\right)\theta_\varepsilon(x)dx = \langle g\rangle \int_\Omega \theta_\varepsilon(x)dx$$

(2.25)

Putting together (2.24) and (2.25), the claim follows. □

2.3 1D Examples: Direct and Inverse Homogenization Problems in Conductivity and Thermo-elasticity

In this subsection we present two basic one-dimensional examples. These examples provide an idea of homogenization proofs in the simplest possible settings.

Example 2.1 (1D Conductivity Problem)

Given $f \in L^2(0, 1)$, let u_ε be the unique solution of

$$\frac{d}{dx}\left[a(x/\varepsilon)\frac{d}{dx}u_\varepsilon(x)\right] = f(x) \text{ in } (0, 1)$$

(2.26)

25

2

2.3 · 1D Examples: Direct and Inverse Homogenization Problems in...

$$u_\varepsilon(0) = 0, \ u_\varepsilon(1) = 0, \tag{2.27}$$

where $0 < \alpha \leq a(y) \leq C$ is a (given) 1−periodic function.

We expect that $u_\varepsilon \to u_0$ in some (weak) sense where u_0 satisfies the following equation:

$$\frac{d}{dx}\left[a_0 \frac{d}{dx}u_0(x)\right] = f(x) \text{ in } (0, 1) \tag{2.28}$$

$$u_0(0) = 0, \ u_0(1) = 0 \tag{2.29}$$

and make a guess that a_0 is constant. This is proved in the next proposition.

Proposition 2.1

Let u_ε be a solution of the problem (2.26)–(2.27). Then

$$u_\varepsilon \rightharpoonup u_0 \text{ in } H_0^1(0, 1) \tag{2.30}$$

where u_0 solves (2.28)–(2.29), and

$$a_0 = \left\langle a^{-1} \right\rangle^{-1}. \tag{2.31}$$

∎

Proof

Step I (Existence of the weak H^1 limit of u_ε). Multiply (2.26) by u_ε, and integrate it over $(0, 1)$ to get (after integrating by parts)

$$\int_0^1 a\left(\frac{x}{\varepsilon}\right)\left|\frac{du_\varepsilon}{dx}\right|^2 dx = -\int_0^1 fudx. \tag{2.32}$$

Using the Cauchy-Shwartz and Friedrichs inequalities we obtain

$$\alpha \left\|\frac{du_\varepsilon}{dx}\right\|_{L^2}^2 \leq \int a\left(\frac{x}{\varepsilon}\right)\left|\frac{du_\varepsilon}{dx}\right|^2 dx \leq \|f\|_{L^2}\|u_\varepsilon\|_{L^2} \leq C\left\|\frac{du_\varepsilon}{dx}\right\|_{L^2}. \tag{2.33}$$

Thus u_ε is uniformly bounded in H^1 and therefore has a weak limit u_0, up to extracting a subsequence. We need to compute this limit. To this end integrate (2.26) in x from 0 to x and introduce:

$$p^\varepsilon(x) := a\left(\frac{x}{\varepsilon}\right)\frac{du^\varepsilon}{dx} = F(x) + c_\varepsilon, \quad F(x) := \int_0^x f(x)dx, \tag{2.34}$$

where c_ε is a constant of integration, depending on ε.

Step II (Computation of the weak limits of p^ε and $\frac{du^\varepsilon}{dx}$). In order to find the limit of p^ε, first compute the limit of c_ε. Divide (2.34) by $a(x/\varepsilon)$, integrate in x from 0 to 1, and use (2.27) to get

$$0 = \int_0^1 \frac{du_\varepsilon}{dx} dx = \int_0^1 a^{-1}\left(\frac{x}{\varepsilon}\right) F(x)dx + c_\varepsilon \int_0^1 a^{-1}\left(\frac{x}{\varepsilon}\right) dx. \tag{2.35}$$

Note that (2.35) also provides a formula for c_ε. Using the Averaging lemma we find

$$\lim_{\varepsilon \to 0} \int_0^1 a^{-1}\left(\frac{x}{\varepsilon}\right) dx = \langle a^{-1}\rangle \text{ and } \lim_{\varepsilon \to 0} \int_0^1 a^{-1}\left(\frac{x}{\varepsilon}\right) F(x)dx = \langle a^{-1}\rangle \int_0^1 F(x)dx. \tag{2.36}$$

Thus, from (2.35) and (2.36) we obtain

$$\lim_{\varepsilon \to 0} c_\varepsilon = -\langle a^{-1}\rangle^{-1}\langle a^{-1}\rangle \int_0^1 F(x)dx = -\int_0^1 F(x)dx. \tag{2.37}$$

Now use (2.34) to compute the limit of p^ε and to establish the *convergence of fluxes*:

$$\text{wlim}_{\varepsilon \to 0}\, p_\varepsilon = F(x) + \lim_{\varepsilon \to 0} c_\varepsilon = F(x) - \int_0^1 F(x)dx =: p_0. \tag{2.38}$$

Exercise 2.3

Equation (2.38) yields $p_0(x) = F(x) - \langle F(x)\rangle$. Derive this representation for p_0 directly from (2.28)–(2.29). ■

Observe that Lemma 2.1 and (2.34) yield

$$\text{wlim}_{\varepsilon \to 0}\, \frac{du^\varepsilon}{dx} = \langle a^{-1}\rangle F(x) + \langle a^{-1}\rangle \lim_{\varepsilon \to 0} c_\varepsilon =: v. \tag{2.39}$$

Step III (Derivation of the effective constitutive equation and homogenization limit). In order to derive the effective constitutive equation, we need to link p_0 and u_0. Since $v \in L^2(0, 1)$ with $\langle v\rangle = 0$ (by (2.37)), the function u_0 defined by

$$u_0(x) = \int_0^x v(s)ds \tag{2.40}$$

belongs to $H_0^1(0, 1)$. Using (2.37), (2.38) and the definition of v by (2.39), we see that the derivative of u_0 satisfies:

$$\frac{du_0}{dx} = \langle a^{-1}\rangle \left[F(x) - \int_0^1 F(x)dx \right] = \langle a^{-1}\rangle p_0. \tag{2.41}$$

Thus we establish the effective constitutive equation for the homogenized flux:

$$p_0 = a_0 \frac{du^0}{dx} \text{ where } a_0 = \langle a^{-1}\rangle^{-1} \tag{2.42}$$

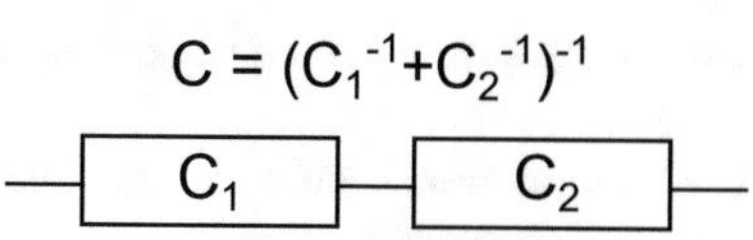

2.3 · 1D Examples: Direct and Inverse Homogenization Problems in...

27

2

Fig. 2.5 Resistors in series

$$C = (C_1^{-1} + C_2^{-1})^{-1}$$

Note that from the third equality of (2.38) and definition of $F(x)$ in (2.34), $\frac{dp_0}{dx} = f(x)$. Thus, differentiation of (2.42) leads to:

$$\frac{d}{dx}\left[a_0 \frac{du_0}{dx}\right] = \frac{d}{dx}\left[\langle a^{-1}\rangle^{-1}\frac{du_0}{dx}\right] = \frac{dp_0}{dx} = f(x), \ x \in (0, 1).$$

Therefore (2.28) is recovered, which completes the proof.

Remark 2.2 Recall the constitutive equation for the problem (2.26):

$$p_\varepsilon = a\left(\frac{x}{\varepsilon}\right)\frac{du_\varepsilon}{dx}, \tag{2.43}$$

where in terms of Hooke's law for elasticity, p_ε represents stress, $a\left(\frac{x}{\varepsilon}\right)$ is stiffness of the material and $\frac{du_\varepsilon}{dx}$ represents strain. In terms of Ohm's law, p_ε represents current, $a\left(\frac{x}{\varepsilon}\right)$ represents conductivity, and $\frac{du_\varepsilon}{dx}$ represents the electric field. In terms of Fourier's law of cooling, p_ε represents heat flux, $a\left(\frac{x}{\varepsilon}\right)$ represents thermal conductivity, and $\frac{du_\varepsilon}{dx}$ is the gradient of temperature.

Remark 2.3 The formula for $a_0 = \langle a^{-1}\rangle^{-1}$ can be understood intuitively in the setting of electrostatics. The conductance C of resistors in series with conductances C_i is depicted on the ■ Figure 2.5. The effective conductance in Proposition 2.1 is analogous to the discrete case of conductance in a series circuit.

Exercise 2.4

Consider the 1D conductivity problem

$$\frac{d}{dx}\left(a\left(\frac{x}{\varepsilon}\right)\frac{d}{dx}u_\varepsilon(x)\right) = f\left(\frac{x}{\varepsilon}\right), \quad x \in \Omega, \tag{2.44}$$

$$u_\varepsilon(0) = u_\varepsilon(1) = 0, \tag{2.45}$$

where $f \in L^2_{per}(0, 1)$ and $0 < \alpha \le a(y) \le \beta$ is a given 1-periodic function. Derive the effective problem for the homogenized limit $u_\varepsilon \to \hat{u}$. Find the value of the homogenized coefficient $\hat{a}$. ■

The following example was suggested by A. Kolpakov and written jointly with him.

Example 2.2 (Inverse homogenization problem for a thermo-elastic one dimensional rod.)

Consider an elastic rod with fine scale microstructure that is ε-periodic, where $\varepsilon > 0$ is a small parameter. Assume that the rod is heated and the external force $f(x/\varepsilon)$ is applied along the rod, and the rod is clamped on its ends. Then deformations of the rod are described by the following thermo-elasticity equation [57]:

$$\frac{d}{dx}\left[\kappa(x/\varepsilon)\left(\frac{d}{dx}u_\varepsilon(x) - \alpha(x/\varepsilon)\right)\right] = f(x/\varepsilon) \quad x \in [0, 1], \tag{2.46}$$

$$u_\varepsilon(0) = u_\varepsilon(1) = 0. \tag{2.47}$$

Here $u_\varepsilon(x)$ is the displacement, $\kappa(y)$ is the Young's modulus, and $\alpha(y)$ is the thermal expansion coefficient; both $\kappa(y)$ and $\alpha(y)$ are 1-periodic.

Exercise 2.5

Similarly to Proposition 2.1 (see also Exercise 2.4), derive the following homogenized equation

$$\frac{d}{dx}\left[\hat{\kappa}\left(\frac{d}{dx}v(x) - \hat{\alpha}\right)\right] = \langle f \rangle \quad x \in [0, 1], \tag{2.48}$$

$$v(0) = v(1) = 0, \tag{2.49}$$

with $\hat{\kappa} = \langle \kappa^{-1} \rangle^{-1}$ and $\hat{\alpha} = \langle \alpha \rangle$, and establish the convergences

$$u_\varepsilon \rightharpoonup v \text{ weakly in } H_0^1(0, 1), \tag{2.50}$$

$$\kappa(x/\varepsilon)\left(\frac{d}{dx}u_\varepsilon - \alpha(x/\varepsilon)\right) \rightharpoonup \hat{\kappa}\left(\frac{d}{dx}v - \hat{\alpha}\right) \text{ weakly in } L^2(0, 1). \tag{2.51}$$

Note that the constant $\hat{\alpha}$ can be omitted in the equation (2.48) but it also appears in the convergence of fluxes and so it is preserved in (2.48) for consistency with (2.51). ∎

The following two questions arise.

1. First, given a specific set of periodic functions $\{(\kappa_j(y), \alpha_j(y))\}_{j \in J}$ corresponding to material properties, describe the corresponding set of effective coefficients $\hat{\kappa}, \hat{\alpha}$. In physical terms, if one is given a library of periodic microstructures, describe the corresponding set of effective properties.

2. Second, what material properties of the constituents are required to design a material with specific (desired) effective properties? Mathematically, given the homogenized coefficients $\hat{\kappa}, \hat{\alpha}$, describe the set of all possible coefficients $(\kappa(y), \alpha(y))$ such that the convergences (2.50)–(2.51) hold.

While issue (1) addresses the *direct homogenization problem*, issue (2) provides an example of an *inverse homogenization problem*, when coefficients describing the microstructure need to be restored from known effective coefficients. In general the inverse problem is ill-posed and typically one can only describe a set of such microstructures. In practical problems one wants to optimize the effective properties with respect to, e.g., cost of design. There is a vast literature on such problems called optimal design of composite materials, when one has a library of available materials and chooses various geometries to achieve required effective properties [4, 34, 63, 75], and references therein.

We illustrate the mathematical essence of this problem using our one dimensional example (2.46)–(2.47). Rewrite the homogenized coefficients

$$\hat{\kappa} = \left(\int_0^1 \frac{1}{\kappa(y)} dy \right)^{-1} , \quad \hat{\alpha} = \int_0^1 \alpha(y) dy. \tag{2.52}$$

When addressing issue (2) the equations in (2.52) is viewed with respect to unknowns $\kappa(y)$, $\alpha(y)$ while $\hat{\kappa}$, $\hat{\alpha}$ are given. Consider the equations in (2.52) and suppose we have, e.g., five types of homogeneous materials, that is the functions $\kappa(y)$ and $\alpha(y)$ take five pairs of values (κ_i, α_i), $i = 1, \ldots, 5$. Then the integrals in (2.52) become *convex combinations*, i.e.,

$$\frac{1}{\hat{\kappa}} = \int_0^1 \frac{1}{\kappa(y)} dy = \sum_{i=1}^5 \int_{A_i} \frac{1}{\kappa_i} dy = \sum_{i=1}^5 \frac{1}{\kappa_i} \beta_i, \quad \hat{\alpha} = \sum_{i=1}^5 \alpha_i \beta_i, \tag{2.53}$$

where A_i is the set where $\kappa(y) = \kappa_i$ (and $\alpha(y) = \alpha_i$) and $\beta_i = |A_i|$. Clearly $\beta_i \geq 0$ and $\sum_{i=1}^5 \beta_i = 1$. The following picture (🔲 Figure 2.6) illustrates the concept of attainable effective coefficients. It shows that any point inside the convex polygon generated by the five points $(1/\kappa_i, \alpha_i)$, $i = 1, \ldots, 5$ in the $(1/\kappa, \alpha)$-plane can be attained whereas any point outside cannot be.

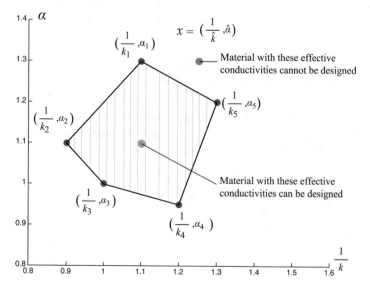

🔲 **Fig. 2.6** The convex polygon of attainable effective coefficients for a given set of five homogeneous materials

We conclude this example by noting that in dimensions two and three homogenization in thermoelastic systems reveals a striking feature of negative effective thermal expansion coefficients [66]. This means that the material shrinks when heated which is counterintuitive, just as the negative Poisson ratio effect to be presented in Example 3.3, ▶ Section 3. Negative effective thermal expansion was obtained for *laminated structures* which are composed of stratified materials whose elastic and thermoelastic properties vary only in one direction. Homogenization of laminated structures leads to solving ODEs and therefore quite often to explicit formulas for effective coefficients [63], similar to explicit formulas in (2.53). Composites with negative effective thermal expansion are obtained by mixing of two or more homogeneous materials (phases) with different thermoelastic properties. When the temperature increases, the usual thermal expansion occurs in each phase. However, since these phases are "bound" together (the composite material is not broken) the macroscopic negative thermal expansion occurs. To explain this phenomenon, suppose that a laminated structure is composed of alternating layers with two vastly different thermal expansion coefficients. A layer with the higher thermal expansion coefficient undergoes expansion deformations in the directions perpendicular to this layer (cross-layer direction) and parallel to this layer (in-plane directions). The in-plane expansion is transmitted through the contact interface to a neighboring layer which is much softer and has much smaller thermal expansion coefficient. Under some conditions on elastic moduli of this layer, it undergoes the contractile deformations in the cross-layer direction due to the Poisson effect (horizontal stretching results in a vertical contraction). Then a proper choice of thickness of the alternating layers leads to the contractile deformations of the composite in the cross-layer direction when heated. As mentioned above this homogenization problem for laminated structure is described by functions of one variable that satisfy ODEs. However, since elasticity is intrinsically a vectorial problem, these functions are vector-valued which leads to technical complications that fall outside the scope of this introductory textbook, see [63] for mathematical description of the negative thermal expansion effect.

2.4 Why is Homogenization Indispensable in Multiscale Problems? Numerical Examples

In this subsection we consider the case study conductivity problem in one and two dimensions, and provide numerical comparison between simulations for the direct and the homogenized problem.

Example 2.3
We first consider the one-dimensional problem:

$$-(\sigma(\frac{x}{\varepsilon})u'_\varepsilon)' = 1, \quad 0 < x < 1, \tag{2.54}$$

$$u_\varepsilon(0) = 0, \quad u_\varepsilon(1) = 0. \tag{2.55}$$

where

$$\sigma(y) = 1 + H(\sin(2\pi(y - 0.5))) = \begin{cases} 1, & 0 < y < 0.5 \\ 2, & 1 > y > 0.5 \end{cases} \quad \text{and } \sigma \text{ is 1-periodic} \qquad (2.56)$$

(H is the Heaviside function). The solution $u_\varepsilon(x)$ is compared to the solution of the homogenized problem

$$-\hat{\sigma} u_0'' = 1, \quad 0 < x < 1, \qquad (2.57)$$

$$u_0(0) = u_0(1) = 0. \qquad (2.58)$$

In this one-dimensional case the effective coefficient $\hat{\sigma}$ is the harmonic mean of σ over the period, that is, $\hat{\sigma} = \dfrac{2}{1/1 + 1/2} = \dfrac{4}{3}$. Thus, the homogenized solution can be easily found analytically:

$$u_0(x) = \frac{3}{8}x(1 - x). \qquad (2.59)$$

The simulation results are presented in ◻ Figure 2.7.

The first comparison shows good agreement between the "exact solution" of (2.54)–(2.55) and the homogenized equation (2.57)–(2.58) for $\varepsilon = 0.1$. The "exact solution" is obtained by solving (2.54)–(2.55) with very fine mesh (size $h = 1/400 = .0025$). The second comparison shows that the direct numerical solution of the two-scale problem (2.54)–(2.55) with mesh size of order ε is very inaccurate and thus one must choose a much smaller mesh size which increases the computational complexity. These simulations show the computational advantage of homogenization even in the simplest 1D problem. ∎

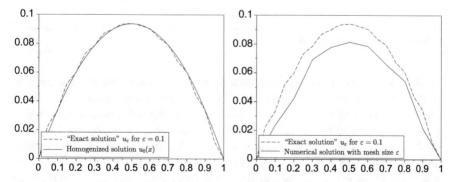

◻ **Fig. 2.7** (left) Comparison between "exact" and homogenized solutions (right) Comparison between "exact" and numerical solution with mesh size $\sim \varepsilon$.

Example 2.4

We next consider the two-dimensional problem:

$$-\text{div}(\sigma(\frac{x}{\varepsilon})\nabla u_\varepsilon) = 1, \quad 0 < x_1, x_2 < 1, \qquad (2.60)$$

$$u_\varepsilon = 0, \quad x \in \partial\Omega \qquad (2.61)$$

where $\sigma(x_1, x_2) = \sigma_1 = 1$ when $(x_1 - 1/2)(x_2 - 1/2) \geq 0$, and $\sigma(x_1, x_2) = \sigma_2 = 5$ when $(x_1 - 1/2)(x_2 - 1/2) < 0$. Function $\sigma(x_1, x_2)$ is extended by periodicity to obtain a $1-$periodic checkerboard pattern in $\mathbb{R}^2$. The periodicity cell in this problem is depicted in ◻ Figure 3.1, ▶ Section 3.

As in the one-dimensional example, we compare the solution u_ε of (2.60)–(2.61) with the homogenized solution u_0 of the following problem:

$$-\hat{\sigma}\Delta u_0 = 1, \quad 0 < x_1, x_2 < 1, \qquad (2.62)$$

$$u_0 = 0, \quad x \in \partial\Omega, \qquad (2.63)$$

where $\hat{\sigma} = \sqrt{5}$ is found by solving the cell problem (2.14)–(2.15). Note that in the case of a periodic checkerboard the cell problem admits an explicit solution $\hat{\sigma} = \sqrt{\sigma_1\sigma_2}$ (see ▶ Section 7).

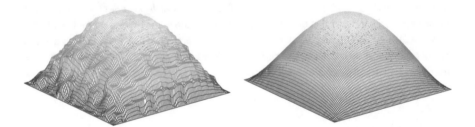

◻ **Fig. 2.8** (left) "Exact solution" of two-scale problem (2.60)–(2.61) for $\varepsilon = 0.1$ and (right) solution of homogenized problem (2.62)–(2.63).

◻ Figure 2.8 shows good agreement between the exact solution of the two-scale problem (2.60)–(2.61) with $\varepsilon = 0.1$ (left) and the solution of the homogenized equation (2.62)–(2.63) (right). The "exact solution" of the two-scale problem (2.60)–(2.61) is obtained by taking a very fine mesh (size $h = 1/128 = 0.0078$). The solution of the homogenized equation (2.62)–(2.63) is obtained using a mesh size $h = 0.04$. The "exact solution" is also compared to the direct numerical solution of (2.60)–(2.61) by using the same mesh size $h = 0.04$ which is less than half of the size of one period ε, see ◻ Figure 2.9, right. Even for such a small mesh, the direct numerical simulation exhibits large deviations from the "exact solution". The plots in ◻ Figure 2.9 show the advantage of the homogenization approach: for the same mesh size the homogenized solution agrees with the "exact solution" quite well while the direct numerical solution is much less accurate. These observations can be quantified by comparing

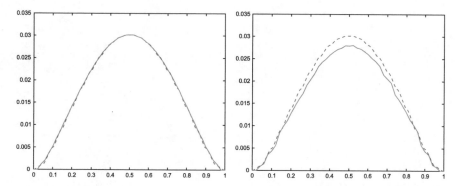

□ **Fig. 2.9** (left) Comparison between "exact solution" of (2.60)–(2.61) (dashed line) and homogenized solution of (2.62)–(2.63) (solid line) and (right) comparison between "exact solution" of (2.60)–(2.61) (dashed line) and direct numerical solution of (2.60)–(2.61) with mesh size $h = 0.04$ (solid line). Both figures depict the cross-section of solutions along the diagonal $x_1 = x_2$.

the numerical percentage error which is in 2D approximately 0.07% for the homogenized solution and 7.4% for direct numerical simulations.   ∎

Both examples show that to achieve reasonable accuracy in direct simulations, the mesh size must be taken much less than the fine scale ε. As the number of nodes in a mesh of size h in 3D is $O(h^3)$, the computational cost becomes prohibitively expensive even for modern computers.

Brief History and Surprising Examples in Homogenization

© Springer Nature Switzerland AG 2018
L. Berlyand, V. Rybalko, *Getting Acquainted with Homogenization and Multiscale*,
Compact Textbooks in Mathematics,
https://doi.org/10.1007/978-3-030-01777-4_3

In this chapter we present a brief overview of the development of homogenization theory including, in particular, several striking examples where mathematical results contrast first intuition.

Homogenization ideas appeared in the works of some of the most prominent scientists. In 1826, Poisson [89] studied the effective conductivity of a composite (mixture between two pure forms of material) consisting of a non-conductive matrix with conducting spherical inclusions. Similar work was done by Faraday in 1839. In 1873, Maxwell [71] (see also [61]) found a formula for effective conductivity of an array of spheres embedded into a matrix with different conductivities ($\sigma_1, \sigma_2 > 0$, $\sigma_1 \neq \sigma_2$). It was similar to our case study (2.5)–(2.6), but he studied a so-called dilute case when each inclusion had a radius r, with $r \ll \ell$, where ℓ was the distance between the inclusions, so that there is almost no interaction between the inclusions. In addition to the large number of inclusions, Maxwell introduced a small parameter $c = \frac{r}{\ell} \to 0$, which corresponds to the volume fraction of inclusions $\phi := \frac{\frac{4}{3}\pi r^3}{\ell^3}$ being small. The cell problem in this setting can be solved asymptotically in the limit $c \to 0$, which is what Poisson and Maxwell did (see [62]). In 1892, Rayleigh [90] gave a mathematically accurate solution of a two-dimensional problem for a periodic array of discs using complex analysis. In 1906, Einstein [46] computed the effective viscosity $\hat{\eta}$ of hard spheres suspended in a Newtonian fluid of viscosity η_0. He considered the dilute case, when the volume fraction $\phi \to 0$ and found the following formula for effective viscosity

$$\hat{\eta} = \eta_0[1 + \frac{5}{2}\phi + O(\phi^2)], \ \phi \to 0$$

All of these results (except Rayleigh's 2D solution) required outstanding physical intuition. However, since problems of heterogeneous media are ubiquitous in various areas in science and engineering, there is a need for a *systematic* mathematical approach (theory) to a broad class of such multiscale problems.

In 1964 three influential papers began the development of such mathematical theory:

1) V. Marchenko and E. Khruslov [70] developed a systematic approach to the Dirichlet problem in a perforated domain (domain with fine grained boundary) that did not require periodicity, but was based on the mathematical notion of capacity and potential theory techniques [69, 99, 100]. In particular, they proved the existence of a homogenization limit for such problems which contains an additional term (also called "strange" term in the literature) due to the influence of perforations.

2) J.B Keller [64] obtained a beautiful explicit formula for the effective conductivity of a checkerboard, where the black squares of the checkerboard have conductivity σ_1 and the white squares have conductivity σ_2 (☐ Figure 3.1). He proved that the effective conductivity satisfies the geometric mean law, $\hat{\sigma} = \sqrt{\sigma_1\sigma_2}$.

3) M. Freidlin [51] considered parabolic PDEs with periodic rapidly oscillating coefficients. He derived the homogenization limit for such problems using probabilistic techniques.

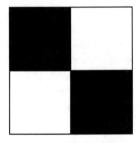

☐ **Fig. 3.1** Periodicity cell in a checkerboard two-phase medium.

In the late 1960s, many significant contributions to homogenization theory were made. A number of fundamental results were obtained by E. De Giorgi, who pioneered the framework of Γ-convergence. Also, the concept of abstract convergence of *differential operators* was developed. F. Murat and L. Tartar [78] introduced the notion of H-convergence of differential operators via the convergence of the resolving operators. For example, consider equations

$$\operatorname{div}\left(\sigma_\varepsilon \nabla u_\varepsilon\right) = f$$

in a bounded domain Ω where the conductivity matrix σ_ε oscillates rapidly as $\varepsilon \to 0$, as in our case study problem. If solutions u_ε converge to the solution of the homogenized equation

$$\operatorname{div}\left(\hat{\sigma} \nabla u\right) = f,$$

for every right-hand side f, the corresponding differential operators are said to be H-convergent.

Next, ideas of periodic homogenization were developed through the work of A. Bensoussan, J. Lions, G. Papanicolaou, O. Oleinik, S. Kozlov, V. Jikov, N. Bakhvalov, I. Babuska, and V. Berdichevsky among others. They developed a systematic approach to periodic homogenization which can be viewed as the "Multiscale Calculus." In the 21st century, significant advancements were made by many authors, e.g., the study of fully nonlinear problems by L. Caffarelli and P.-L. Lions. Finally we mention that the term "homogenization" was coined by I. Babuska [10].

Recall that goal of the general approach is to predict the expected homogenization limit, and then to justify it by proving some type of convergence as $\varepsilon \to 0$. In the case study problem (2.5)–(2.6), taking the limit as $\varepsilon \to 0$ yields an equation of the same type, and only the coefficients change. The following examples show that the limiting (homogenized) equation may feature different and quite surprising behavior.

Example 3.1 ("Strange Term Coming from Nowhere")

Consider the following problem of a dilute limit where the inclusions are small and far apart, so that their total volume fraction goes to zero:

$$-\Delta u^\varepsilon = f, \; x \in \Omega \setminus \cup_{i=1}^{N_\varepsilon} B_i^\varepsilon \tag{3.1}$$

$$u^\varepsilon = 0, \; x \in \overline{B_i^\varepsilon} \tag{3.2}$$

$$u^\varepsilon = 0, \; \text{on } \partial\Omega, \tag{3.3}$$

where the B_i^ε ($i = 1 \ldots N_\varepsilon \to \infty$ as $\varepsilon \to 0$) are small, identical inclusions. This problem describes, in particular, equilibrium of an elastic membrane pinned by many small nails. The domain, $\Omega_\varepsilon := \Omega \setminus \cup_{i=1}^{N_\varepsilon} B_i^\varepsilon$ is said to have a fine-grained boundary [70] (◻ Figure 3.2).

◻ **Fig. 3.2** Dilute system: inclusions are tiny and far apart.

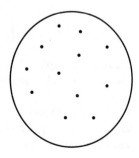

It was shown [38, 70] that the limiting homogenized problem is of the form

$$-\Delta u_0 + \hat{g}(x)u_0 = f, \; x \in \Omega, \tag{3.4}$$

thus containing an additional term $\hat{g}(x)u_0$. In a more complicated example, a boundary value problem for PDEs can lead to an integro-differential equation in the homogenization limit [69].

■

The development of the mathematical approach in [70] was motivated by joint work with physicists to model wave reflection on diffraction gratings (closely packed, very narrow grooves engraved into the grating's surface, see ◻ Figure 3.3). An attempt to directly compute the solution of this problem failed because of the high computational complexity due to the large number of narrow grooves. Then the homogenization approach helped to resolve this difficulty. This example illustrates our previous statement that homogenization theory was born out of computational fight with fine scales. While the development of homogenization theory resulted in many beautiful ideas and approaches in analysis, PDEs, and calculus of variations, the ultimate goal has been to provide an efficient computational tool for applied multiscale problems.

◻ **Fig. 3.3** Diffraction gratings

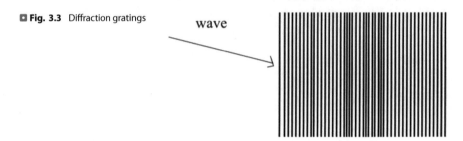

The following examples illustrate an application of homogenization theory in geometry and mechanics.

Recall that many PDE problems can be formulated in a variational form, when solutions minimize a certain energy functional. Next we introduce a two-scale variational problem for functions with rapidly oscillating coefficients,

$$\min_{u^\varepsilon \in K} F^\varepsilon[u^\varepsilon] = \min_{u^\varepsilon \in K} \int_\Omega f\left(x, \frac{x}{\varepsilon}, u, \nabla u^\varepsilon\right) dx \qquad (3.5)$$

where the integrand $f(y, z)$, called the *Lagrangian*, is periodic in $y = \frac{x}{\varepsilon}$, K is a class of admissible functions. In particular, taking the Lagrangian $f(x, y, u, \xi) = \sum_{ij} \frac{1}{2}\sigma_{ij}(y)\xi_i\xi_j - g(x)u$, we recover our case study problem (2.5)–(2.6) for $K := H_0^1$. In this case (3.5) is simply a reformulation of (2.5)–(2.6) in a variational form. It will be shown that the asymptotic behavior of minimizers of (3.5) as $\varepsilon \to 0$ is described by a variational problem of the form

$$\min_{u \in \hat{K}} \hat{F}(u) = \min_{u \in \hat{K}} \int_\Omega \hat{f}(u, \nabla u) dx$$

where $\hat{f}$ is the homogenized Lagrangian. For the case study problem, $\hat{f}(x, u, \nabla) = \frac{1}{2}\hat{\sigma}|\nabla u|^2 - g(x)u$.

Example 3.2 (A "Hilly Landscape" Problem)

Recall the notion of *Riemannian manifold* and *Riemannian metric*. A Riemann manifold, M, is a space which locally "is similar" to $\mathbb{R}^n$ and has an inner product on the tangent space at each point of the manifold. It is defined via a tensor often denoted $g = (g_{ik})_{i,k=1..n}$ and called the *Riemann metric*. This metric generalizes the notion of geodesic distance in abstract spaces. For example, in the Euclidean metric the arc length of a curve $\gamma : [0, 1] \to \mathbb{R}^2$ is defined by

$$s = \int_0^1 \sqrt{(dx_1)^2 + (dx_2)^2},$$ (3.6)

written infinitesimally as $ds^2 = \sum(dx_i)^2$. In the general ds^2 is defined via the Riemann metric by

$$ds^2 = \sum g_{ik}(x)dx_i dx_k.$$ (3.7)

In (3.7) g_{ik} provides anisotropy and inhomogeneity if compared with (3.6). Consider *geodesics* for an *oscillating* Riemannian metric. Assume that $M = \mathbb{R}^n$ is endowed with an ε-periodic Riemannian metric $\sigma^\varepsilon(x) = \sigma(x/\varepsilon)$. Recall that a geodesic between two points x_1 and x_2 is the shortest path between x_1 and x_2 on the surface with respect to the (in this case) Riemannian metric. This amounts to minimization of

$$d_\varepsilon^2(x_1, x_2) = \inf \left\{ \int_0^1 \sum_{i,j} \sigma_{ij}\left(\frac{u}{\varepsilon}\right) u_i' u_j' ds, \; u(0) = x_1, \; u(1) = x_2 \right\}$$ (3.8)

among the curves $u = (u_1, u_2) : [0, 1] \to M$ connecting points x_1 and x_2. Exploring this metric, one can think of "periodic hills," as depicted in ◻ Figure 3.4. Since "hills" are very expensive to go through with respect to the Riemannian metric, it requires less Riemannian distance to go around them.

◻ **Fig. 3.4** Oscillating Riemannian metric, black regions correspond to "hills" where the metric tensor is high.

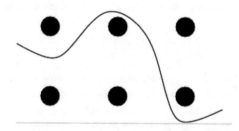

The limiting functional of (3.8) as $\varepsilon \to 0$ is

$$\min \left\{ \int_0^1 \psi(u')ds : \; u(0) = x_1, \; u(1) = x_2 \right\}$$ (3.9)

where ψ is not quadratic. A surprising result is that (3.9) is not related to any Riemannian metric, except for some trivial cases. The metric ψ is known as a *Finsler* metric in geometry. The limiting metric is obtained in [28] by Γ-convergence techniques (introduced in ▶ Section 9), see also [29] for a geometric approach to this problem.   ∎

Example 3.3 (A Surprising Physical Prediction of Homogenization Theory)

Recall that an isotropic, elastic body is described by the so-called Lamé-Navier system of PDEs

$$-\text{div}\,\sigma\,(u(x)) = f(x), \quad x \in \Omega \tag{3.10}$$

$$\sigma_{ii}(u) = \lambda\text{div}\,u + 2\mu\frac{\partial u_i}{\partial x_i}, \quad \sigma_{ij} = \mu\left(\frac{\partial u_i}{\partial x_j} + \frac{\partial u_j}{\partial u_i}\right) \quad (i \neq j), \tag{3.11}$$

where λ, μ are *Lamé's constants*, and u is the displacement. For a general anisotropic elastic medium Hooke's law has the form:

$$\sigma_{ij} = c_{ijk\ell}\epsilon_{k\ell}, \tag{3.12}$$

where we adopt the Einstein summation convention (i.e., sum over repeated indices). The strain tensor, ϵ, has the form

$$\epsilon_{k\ell} = \frac{1}{2}\left(\frac{\partial u_k}{\partial x_\ell} + \frac{\partial u_\ell}{\partial x_k}\right) \tag{3.13}$$

and for isotropic media the elasticity (stiffness) tensor c simplifies as follows:

$$c_{iiii} = \lambda + 2\mu, \quad c_{iikk} = \lambda, \quad c_{ikik} = 2\mu, \quad (i \neq k), \tag{3.14}$$

with $c_{ijk\ell} = 0$ otherwise. Analogously to the Dirichlet principle for scalar problems (▶ Section 1), the solution u of (3.10) minimizes the energy functional

$$\frac{1}{2}\int_\Omega W(u, u)dx - \int_\Omega (f, u)dx. \tag{3.15}$$

If λ, μ are periodic and piecewise constant functions, then one may apply homogenization theory. These coefficients describe the physical properties of the medium: in particular, the coefficient $\nu = \nu(\lambda, \mu)$ is called the *Poisson ratio*, and quantifies the ratio between the vertical shortening ϵ_{11} and horizontal elongation ϵ_{22}:

$$\nu := -\frac{\epsilon_{22}}{\epsilon_{11}}. \tag{3.16}$$

As expected intuitively, normal materials compress vertically when elongated horizontally, which corresponds to positive Poisson ratio, $\nu > 0$ (◘ Figure 3.5). Homogenization theory predicted materials with negative Poisson ratio, that is, materials which fatten vertically when elongated horizontally (■ Figure 3.6). Such models were developed by mixing very stiff and very soft materials. In particular, these models had the property that the constituent materials had positive Poisson ratio and the homogenized material had zero [18] or negative [19, 65, 74] effective Poisson ratio. Further, their use was proposed in the design of transducers [96]. ■

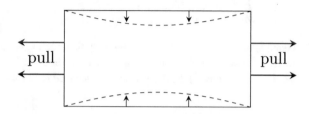

■ **Fig. 3.5** A medium with positive Poisson ratio before and after horizontal pulling

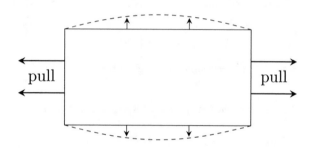

■ **Fig. 3.6** A medium with negative Poisson ratio before and after horizontal pulling

Example 3.4 (Development of New Scales)

Consider the following family of 1D functionals parameterized by the small parameter ε:

$$I_\varepsilon[v] = \int_0^1 (\varepsilon v''(x)^2 + W(v'(x)) + a(x)v(x)^2)dx \tag{3.17}$$

where

$$W(y) = (1 - y^2)^2 \tag{3.18}$$

is a *double-well potential*. When $\varepsilon > 0$ becomes small minimizers of (3.17) exhibit quite nontrivial behavior, developing a new scale of periodic oscillations. In order to explain the nature of these oscillations consider two particular minimization problems for (3.17) when $\varepsilon = 0$.

Case I: $\varepsilon = 0, a(x) = 0$. In this case minimization of (3.17) becomes

$$\min \int_0^1 ((v')^2 - 1)^2 dx. \tag{3.19}$$

It is immediate that the functional (3.19) is bounded below by 0 and therefore any v for which almost everywhere $|v'| = 1$ is a minimizer. These are so-called *sawtooth functions*. Note that there are multiple solutions to problem (3.19) as any sawtooth function is a minimizer.

Case II: $\varepsilon = 0, a(x) = a > 0$. We now seek functions which minimize

$$\min \int_0^1 \left(av^2 + ((v')^2 - 1)^2 \right) dx. \tag{3.20}$$

Again note that (3.20) is bounded below by 0. Since the term av^2 forces $v(x)$ to be small in the L^2 norm, we consider a sequence v_n so that the integral becomes closer to zero (decreasing amplitude). Such a sequence is depicted in ◘ Figure 3.7.

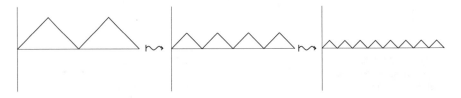

◘ **Fig. 3.7** Sawtooth functions converge to zero

The values of the functional on this sequence converge to zero:

$$I_0[v_n] \to 0 \text{ as } n \to \infty. \tag{3.21}$$

However, observe that there is no minimizer, v, such that $I_0[v] = 0$. Indeed, if $I_0[v] = 0$, then $\|v\|_{L^2} = 0$ so $v \equiv 0$. If we plug $v = 0$ into $I_0[v]$, the term $((v')^2 - 1)^2$ becomes 1 and $I_0[v] = 1$. Thus $v = 0$ is not a minimizer. That is, $v_n \rightrightarrows 0$ (converges uniformly) and $I_0[v_n] \to 0$, yet there is no function v with $I_0[v] = 0$.

Alberti and Müeller [2] studied the functional $I_\varepsilon[v]$ for $\varepsilon > 0$ and showed the following surprising result:

Theorem 3.1 ([2])

Suppose $a > 0$ is a constant. Then for sufficiently small $\varepsilon > 0$ all minimizers of (3.17) among 1-periodic functions have the fundamental period (the smallest period)

$$p_\varepsilon = L_0 a^{-1/3} \varepsilon^{1/3} + O(\varepsilon^{2/3}) \tag{3.22}$$

where $L_0 = \left(96 \int_0^1 \sqrt{W(x)} dx \right)^{1/3}$. ∎

Thus the period p_ε is of order $\varepsilon^{1/3}$. That is, the new scale $\varepsilon^{1/3}$ arises in the solution. Since this period was not present in the original functional (3.17), this example stands in the strong contrast with the case study problem (2.5)–(2.6), in which both the problem and the solution have the same two scales, $O(1)$ and $O(\varepsilon)$. The small parameter (scale) ε in problem (3.17) is present as a coefficient in the first term. The fact that solutions develop a new scale is surprising and demonstrates the utility and necessity of multiscale analysis (❑ Figure 3.8).

❑ **Fig. 3.8** Graph of derivative of minimizer of $I_\varepsilon[v]$

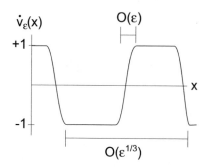

This example is a *singularly perturbed* problem (see also the Appendix). We conclude its study by mentioning three important phenomena that are present in this example and appear in a wide range of asymptotic problems:

1. The overall structure of minimizers is determined by lower order terms of derivatives. That is, the terms of the energy containing v' and v give rise to the large scale sawtooth structures.

2. The minimizers may not be unique in general due to the fact that the energy functional is nonconvex. The novel features due to nonconvexity of the variational problems are presented in ▶ Sections 9 and 11.

3. The highest order term $\varepsilon v''$ provides regularization. That is, the minimizers are smooth and do not contain corners.

Formal Two-Scale Asymptotic Expansions and the Corrector Problem

© Springer Nature Switzerland AG 2018
L. Berlyand, V. Rybalko, *Getting Acquainted with Homogenization and Multiscale*,
Compact Textbooks in Mathematics,
https://doi.org/10.1007/978-3-030-01777-4_4

In this chapter we construct a formal asymptotic expansion of the solution of the case study problem (2.5)–(2.6), repeated below for readers convenience,

$$-\text{div}\left[\sigma(\frac{x}{\varepsilon})\nabla u_\varepsilon\right] = g(x), \ x \in \Omega \tag{4.1}$$

$$u_\varepsilon = 0, \ x \in \partial\Omega, \tag{4.2}$$

by performing asymptotic analysis in the small parameter ε. Asymptotic analysis is a very important tool in the study of applied problems modeled by differential equations such as multiscale problems. It often allows one to drastically reduce computational complexity and sometimes (rarely) leads to explicit analytical solutions.

First, we describe the concept of a formal asymptotic expansion introduced by H. Poincaré.

Definition 4.1 ———

The *formal asymptotic expansion* of a function $f(x, \varepsilon)$, depending on a small parameter ε, is an asymptotic series of the form

$$f(x, \varepsilon) \sim \sum_k \varepsilon^k f_k(x, \varepsilon)$$

(continued)

Definition 4.1 (continued)

such that for any N, there exists N_0 such that $\forall m \geq N_0$, the following formula holds:

$$f(x, \varepsilon) - \sum_{k=1}^{m} \varepsilon^k f_k(x, \varepsilon) = O(\varepsilon^N) \qquad (4.3)$$

where $O(\varepsilon^N)$ is the *discrepancy*.

Note that convergence of the asymptotic series is not required, which is why these series are called *formal*.

Another important idea about the form of the asymptotic expansion is also due to H. Poincaré. In his studies of ODEs of celestial mechanics he introduced two types of variables:

1) The fast variable, $\dfrac{x}{\varepsilon}$, where the change is on the order of ε;

2) The slow variable, x

(e.g., the rotation of the planet around the sun is much slower than its rotation around itself) and proposed to treat them as *independent variables*. This idea simplifies analysis of many asymptotic problems. In particular, in the 1970s this concept of asymptotic expansions with fast and slow variables was successfully applied to PDEs with rapidly oscillating coefficients, which took a significant effort, see, e.g., [15, 16, 93].

Let us return now to our case study example (4.1)–(4.2). Bearing in mind the idea of "fast" and "slow" variables, the following ansatz (i.e., educated guess) is proposed for the solution u_ε:

$$u_\varepsilon(x) = u_0(x, \frac{x}{\varepsilon}) + \varepsilon u_1(x, \frac{x}{\varepsilon}) + \varepsilon^2 u_2(x, \frac{x}{\varepsilon}) + ..., \qquad (4.4)$$

where each function $u_i(x, y)$ is assumed Π−periodic in the "fast" variable $y = \frac{x}{\varepsilon}$. Now substitute (4.4) into equation (2.5) and use the chain rule

$$\nabla u_i(x, \frac{x}{\varepsilon}) = \left(\nabla_x u_i(x, y) + \frac{1}{\varepsilon} \nabla_y u_i(x, y) \right) \Big|_{y=\frac{x}{\varepsilon}}. \qquad (4.5)$$

Grouping terms with like powers of ε yields the following (coupled) equations:

$$\varepsilon^{-2} : L_0 u_0(x, y) = 0 \qquad (4.6)$$

$$\varepsilon^{-1} : L_0 u_1(x, y) + L_1 u_0(x, y) = 0 \qquad (4.7)$$

$$\varepsilon^0 : L_0 u_2 + L_1 u_1 + L_2 u_0 = g(x), \qquad (4.8)$$

47

4

Chapter 4 • Formal Two-Scale Asymptotic Expansions and the Corrector...

where

$$L_0 := -\text{div}_y(\sigma(y)\nabla_y \cdot) \tag{4.9}$$

$$L_1 := -\text{div}_y(\sigma(y)\nabla_x \cdot) - \text{div}_x(\sigma(y)\nabla_y \cdot) \tag{4.10}$$

$$L_2 := -\text{div}_x(\sigma(y)\nabla_x \cdot). \tag{4.11}$$

Since x and y are considered as independent variables, equation (4.6) is a PDE in y-variable with Π-periodicity boundary condition while x is thought of as a parameter. The following lemma implies that u_0 does not depend on the variable y.

🛈 Lemma 4.1 **(Fredholm Alternative for Periodic Elliptic PDEs, [4] Lemma 1.3.21, cf. [49], Sec 6.2.3)** *Assume that $f(y)$ is a Π-periodic function and $f \in L^2(\Pi)$. Then there exists a unique (up to a constant) Π−periodic (with first derivatives) solution $v(y)$ of the problem*

$$-\text{div}[\sigma(y)\nabla v(y)] = f(y), \ y \in \Pi \tag{4.12}$$

if and only if

$$\int_\Pi f(y)dy = 0. \tag{4.13}$$

It is clear that $u_0(x, y) = u_0(x)$ is a solution to (4.6) while Lemma 4.1 implies that this solution is unique (up to a constant that depends on x only). Observe that the equality $\nabla_y u_0(x, y) = 0$ can be established simply by multiplying (4.6) by $u_0(x, y)$ and integrating by parts over Π and using periodic boundary conditions. This shows that the leading term in the asymptotic expansion (4.4) does not depend on the fast variable, that is the homogenization limit does not contain the fine scale. Note that, up to this point, boundary conditions of (4.1)–(4.2) have not been incorporated in the formal expansion (4.4). In order to satisfy the boundary condition (2.6), we impose that $u_0 = 0$ on $\partial\Omega$.

Now rewrite (4.7) as

$$L_0 u_1(x, y) = \text{div}_y[\sigma(y)\nabla_x u_0(x)] = \sum_{j=1}^n \partial_{x_j} u_0(x)\text{div}_y[\sigma(y)\mathbf{e}_j]. \tag{4.14}$$

to see readily that $u_1(x, y)$ depends on $\nabla_x u_0(x)$ linearly. Namely, due to the superposition principle $u_1(x, y)$ is a linear combination of basic solutions $\chi_i(y)$

$$u_1(x, y) = \sum_{j=1}^n \chi_j(y)\partial_{x_j} u_0(x), \tag{4.15}$$

where $\chi_i(y)$ satisfy the equation $L_0\chi_j = \text{div}_y\left(\sigma(y)e_j\right)$ which can be rewritten as follows:

$$\text{div}_y\left(\sigma(y)[\nabla\chi_j(y) + e_j]\right) = 0, \ y \in \Pi \tag{4.16}$$

subject to Π-periodicity condition on the boundary. Note that (4.16) is precisely the cell problem (2.14)–(2.15). What remains is to observe that the homogenized equation from Theorem 2.1 is nothing but the solvability condition for (4.8). To this end rewrite (4.8) in the form

$$L_0u_2 = g(x) - L_1u_1 - L_2u_0, \tag{4.17}$$

and substitute representation (4.15) for $u_1(x, y)$. Recall that x is regarded as a parameter. Then Lemma 4.1 (Fredholm alternative) implies that the solution u_2 exists if and only if

$$\int_\Pi g(x)dy = \sum_{j=1}^n \int_\Pi L_1\left(\partial_{x_j}u_0(x)\chi_j(y)\right)dy + \int_\Pi L_2u_0dy.$$

Since $g(x)$ is constant in y we have $\int_\Pi g(x)dy = g(x)|\Pi|$ and after rearranging terms:

$$|\Pi|g(x) = -\sum_{j=1}^n \int_\Pi \text{div}_x\left(\sigma(y)(\nabla_y\chi_j(y) + e_j)\partial_{x_j}u_0(x)\right)dy$$

$$-\sum_{j=1}^n \int_\Pi \text{div}_y(\sigma(y)\chi_j(y)\nabla_x\partial_{x_j}u_0(x))dy. \tag{4.18}$$

The divergence theorem and the periodicity of $\sigma(y)$ and $\chi_j(y)$ imply that the last term in (4.18) is zero. Let $\overline{\chi} = (\chi_1, ..., \chi_n)$ denote the vector of solutions of the cell problems. Then we can write (4.18) in vectorial form

$$g(x) = -\text{div}_x\left(\frac{1}{|\Pi|}\int_\Pi \sigma(y)(\nabla\overline{\chi}(y) + I)dy\nabla_x u_0(x)\right), \tag{4.19}$$

where I is the identity tensor. Now, introduce

$$\hat{\sigma} := \frac{1}{|\Pi|}\int_\Pi \sigma(y)(\nabla\overline{\chi}(y) + I)dy, \tag{4.20}$$

or, componentwise:

$$\hat{\sigma}_{ij} = \frac{1}{|\Pi|}\int_\Pi e_i \cdot \sigma(y)[e_j + \nabla_y\chi_j(y)]. \tag{4.21}$$

49

4

Chapter 4 • Formal Two-Scale Asymptotic Expansions and the Corrector...

Thus, we get that the solvability condition is

$$-\text{div}_x[\hat{\sigma} \nabla_x u_0] = g(x). \tag{4.22}$$

The expression (4.21) is equivalent to (2.13) in Theorem 2.1 which one can verify by solving the exercise below.

Exercise 4.1

a) Define the effective conductivity tensor in the following two ways:

$$\hat{\sigma}_{ij}^{(1)} = \frac{1}{|\Pi|} \int_\Pi (\sigma(y)(\nabla \chi_i + \mathbf{e}_i)) \cdot (\nabla \chi_j + \mathbf{e}_j) dy \tag{4.23}$$

$$\hat{\sigma}_{ij}^{(2)} = \frac{1}{|\Pi|} \int_\Pi (\sigma(y)(\nabla \chi_i + \mathbf{e}_i)) \cdot \mathbf{e}_j dy \tag{4.24}$$

where χ_i solves the cell problem:

$$-\text{div}[\sigma(y)[\nabla \chi_i + \mathbf{e}_i]] = 0, \; y \in \Pi \tag{4.25}$$

$$\chi_i \in H^1_{per}(\Pi). \tag{4.26}$$

Prove that $\hat{\sigma}^{(1)} = \hat{\sigma}^{(2)}$.

b)[1] More generally, prove the equivalence of the following definitions of $\hat{\sigma}$:
▬ *Flux-Form* (see [62]):

$$\hat{\sigma}\xi = \langle \sigma(y)(\nabla \chi_\xi + \xi) \rangle \tag{4.27}$$

where $\chi_\xi \in H^1_{per}(\Pi)$ with $\langle \chi_\xi \rangle = 0$ and $\tag{4.28}$

$$\text{div}_y(\sigma(y)[\nabla \chi_\xi(y) + \xi]) = 0 \tag{4.29}$$

where $\xi \in \mathbb{R}^n$ is an arbitrary constant vector.
▬ *Variational or Energy Form*:

$$\xi^T \hat{\sigma} \xi = \inf_{v \in H^1_{per}(\Pi)} \langle (\xi + \nabla v)^T \sigma(y)(\xi + \nabla v) \rangle \tag{4.30}$$

where $\sigma(y) = \{\sigma_{ij}(y)\}$, $\hat{\sigma} := \{\hat{\sigma}_{ij}\}$ and $\langle \cdot \rangle := \frac{1}{\Pi} \int_\Pi \cdot \, dy$.

∎

---

[1] This part requires some knowledge on calculus of variations, see ▶ Section 1.5.1 and the proof of the Dirichlet principle in [49] for more details.

Observe that the flux definition states that the homogenized flux is equal to the average flux over the periodicity cell and the energy definition states that the energy of the homogeneous, effective medium is equal to the energy of the heterogeneous, periodic media,

Thus the formal asymptotic expansions lead to the homogenization result in Theorem 2.1. However, these arguments do not provide a rigorous proof, for example the convergence of the series (4.4) is not established. The proof can be done in several ways, but all of them first require developing new tools of "multiscale calculus" which will be done in ▶ Sections 5, 6, and 9.

Recall from Exercise 2.1 that we cannot expect strong L^2 convergence of $\nabla u_\varepsilon \rightarrow \nabla u_0$. This observation led to what has become known as the "corrector" problem. Intuitively the issue is that the function $u_\varepsilon(x)$ has both fine (order ε)) and coarse (order 1) scales and so approximation of ∇u_ε by ∇u_0 fails since u_0 has no fine scale. Thus, we need to improve the approximation beyond the scope of Theorem 2.1. Instead of considering only the principal term in the asymptotic expansion of u_ε we also include the second term $\varepsilon u_1(x, x/\varepsilon)$, defined by means of $u_0(x)$ and $\overline{\chi}(y)$, and u_1 has fine oscillations unlike $u_0(x)$. Set

$$\hat{u}_\varepsilon(x) = u_0(x) + \varepsilon u_1(x, x/\varepsilon), \text{ where } u_1(x, y) = \nabla u_0(x) \cdot \overline{\chi}(y). \tag{4.31}$$

Theorem 4.1
If the conditions of Theorem 2.1 hold and, in addition $\sigma(y) \in C_{per}(\Pi; \mathbb{R}^{n \times n})$, then

$$\|u_\varepsilon(x) - \hat{u}_\varepsilon(x)\|_{H^1(\Omega)} \leq C\sqrt{\varepsilon}. \tag{4.32}$$

Exercise 4.2
Assume that σ and u_0 are C^2-smooth. Substitute $\hat{u}_\varepsilon(x)$ given by (4.31) into (2.5) and calculate the resultant discrepancy. Show that the discrepancy converges to zero weakly in $L^2(\Omega)$ but does not converge to zero strongly in $L^2(\Omega)$. Hint: Use Lemma 2.1 (Averaging Lemma). Compare your solution with Step 2 of the proof of Theorem 2.1 in ▶ Section 5.2.

∎

As a remark, we note the initial surprise of the rate of convergence, $\sqrt{\varepsilon}$, instead of the expected $O(\varepsilon)$. This stems from the fact that although $u_0(x) = 0$ on the boundary $\partial\Omega$, $u_1(x, x/\varepsilon)$ does *not* satisfy the boundary condition. It leads to the so-called *boundary layer effects* and hence $O(\sqrt{\varepsilon})$ convergence. Boundary layer effects are very important in various applied problems. For example, fluid in a channel behaves in a special way near walls. Heuristically, there are two regimes of flow: in a small (size ε) layer near walls (when interaction between fluid and walls is essential) and away from walls (when

51

4

Chapter 4 • Formal Two-Scale Asymptotic Expansions and the Corrector...

the fluid "forgets" about the boundary). This concept is explained in further detail in Appendix A using simple 1-dimensional examples of singular and regular perturbations.

We now describe the main difficulty in the proof of Theorem 2.1. Recall the weak formulation of the case study problem (2.5)–(2.6). For a given $g(x) \in L^2(\Omega)$, find $u_\varepsilon \in H_0^1(\Omega)$ that satisfies

$$\int_\Omega \sigma\left(\frac{x}{\varepsilon}\right) \nabla u_\varepsilon \nabla \phi(x) dx = \int_\Omega g(x)\phi(x)dx, \quad \forall \phi \in H_0^1(\Omega) \tag{4.33}$$

It is natural to try to take the limit in this equation as $\varepsilon \to 0$. We already know by (2.8) that

$$u_\varepsilon \rightharpoonup u_0 \text{ weakly in } H^1(\Omega), \tag{4.34}$$

up to a subsequence. Thus, we have

$$\nabla u_\varepsilon \rightharpoonup \nabla u_0 \text{ weakly in } L^2(\Omega). \tag{4.35}$$

Further, we know that (by Averaging Lemma 2.1)

$$\sigma\left(\frac{x}{\varepsilon}\right) \rightharpoonup \langle\sigma\rangle \text{ weakly in } L^2(\Omega). \tag{4.36}$$

With the weak convergences (4.35) and (4.36) in mind, we may tempted to be optimistic about the convergence of the products in (4.33), and simply deduce that the homogenized conductivity $\hat{\sigma} = \langle\sigma\rangle$. However, this is not the case, as weak convergence of functions does not imply that their products converge weakly to the product of their weak limits (see Exercise 1.1, and 1D case above).

There are several ways of resolving this issue, each leading to the development of different homogenization techniques which can be applied to a wide variety of linear and nonlinear PDEs. We mention here three approaches, the first two will be described in detail below.

1. *Compensated Compactness and Oscillating Test-functions.* The main tool of this method is the Div-Curl Lemma (which itself has wide application). The general idea is to place additional constraints on oscillating sequences that act to "compensate" for the lack of compactness so that the product of weakly convergent sequences does converge to the product of the weak limits.

2. *Two-scale Convergence.* The idea of two-scale convergence is to adjust the notion of weak convergence to functions that contain fast and slow oscillations.

3. *Direct Justification through A Priori Estimates on the Discrepancy.* The goal is to directly justify the two-scale formal expansion by estimating the discrepancies via elliptic estimates. By discrepancy, we mean the following. If we have that

$$\hat{u}^\varepsilon = u_0(x) + \varepsilon u_1\left(x, \frac{x}{\varepsilon}\right) \tag{4.37}$$

then

$$\|u^\varepsilon(x) - u_0(x) - \varepsilon u_1\left(x, \frac{x}{\varepsilon}\right)\|_{H^1(\Omega)} = O(\varepsilon^\gamma), \quad \gamma > 0 \tag{4.38}$$

The right-hand side of (4.38), which is referred to as the discrepancy, would be equal to 0 if (4.37) were an exact solution. This method always works but can become very cumbersome; even for linear but vectorial problems that arise in elasticity theory or Stokes equation for fluids.

ⓘ Remark 4.1 Our model problem (2.5)–(2.6) exhibits several key features of so-called singularly perturbed problems (see Appendix) which play an important role in many applications. First, solutions to singularly perturbed problems develop "fast" scales (i.e., $\frac{x}{\varepsilon}$) in addition to "slow" scales (i.e., x). In contrast, solutions to regularly perturbed problems contain only "slow scales." Second, in regularly perturbed problems one can pass to the limit by simply setting $\varepsilon = 0$ in the original problem whereas in singularly perturbed problems one has to guess the asymptotic ansatz which could be highly nontrivial and problem dependent.

Compensated Compactness and Oscillating Test Functions

© Springer Nature Switzerland AG 2018
L. Berlyand, V. Rybalko, *Getting Acquainted with Homogenization and Multiscale,*
Compact Textbooks in Mathematics,
https://doi.org/10.1007/978-3-030-01777-4_5

5.1 Div-Curl Lemma

In this subsection we describe the so-called Div-Curl Lemma due to F. Murat and L. Tartar [77, 79].

Recall that for any vector-valued function $u \in L^2(\Omega)$, one can define $\operatorname{div} u \in H^{-1}(\Omega)$ by the formula

$$(\operatorname{div} u, \phi) = -\int_{\Omega} u \cdot \nabla \phi \, dx, \quad \forall \phi \in H_0^1(\Omega) \tag{5.1}$$

We can also define $\operatorname{curl}(u)$, where each entry of the curl matrix is in $H^{-1}(\Omega)$ and is defined by

$$(\operatorname{curl}(u), \phi)_{ij} = -\int_{\Omega} \left(u_i \frac{\partial \phi}{\partial x_j} - u_j \frac{\partial \phi}{\partial x_i} \right) dx, \quad \forall \phi \in H_0^1(\Omega) \tag{5.2}$$

In particular, for $n = 3$

$$\operatorname{curl}(u) = ((\operatorname{curl}(u))_{23}, (\operatorname{curl}(u))_{13}, (\operatorname{curl}(u))_{12}). \tag{5.3}$$

🛈 **Lemma 5.1 (Div-Curl Lemma)** *Let P_ε, P_0, V_ε, and V_0 be vector fields in $L^2(\Omega)$ such that*

$$P_\varepsilon \rightharpoonup P_0, \quad V_\varepsilon \rightharpoonup V_0 \ \text{ in } L^2(\Omega) \text{ as } \varepsilon \to 0. \tag{5.4}$$

If in addition

$$\operatorname{div} P_\varepsilon \to \operatorname{div} P_0 \text{ in } H^{-1}(\Omega), \quad \text{and curl } V_\varepsilon = 0, \tag{5.5}$$

then

$$P_\varepsilon V_\varepsilon \rightharpoonup P_0 V_0 \text{ in } \mathscr{D}'(\Omega) = [C_0^\infty(\Omega)]'. \tag{5.6}$$

Recall that the convergence as distributions (in $\mathscr{D}'(\Omega)$) in (5.6) means that $\forall \phi \in C_0^\infty(\Omega)$,

$$\int_\Omega P_\varepsilon V_\varepsilon \phi \, dx \to \int_\Omega P_0 V_0 \, \phi \, dx.$$

ⓘ **Remark 5.1** The name compensated compactness comes from the fact that the additional properties (5.5) compensate for the lack of strong convergence of the factors in the product which in general is needed for passing to weak limits in the product.

Proof

Step 1: Reduction to $P_0 = 0$, $V_0 = 0$. We have

$$\int_\Omega P_\varepsilon V_\varepsilon \phi \, dx = \int_\Omega (P_\varepsilon - P_0)(V_\varepsilon - V_0)\phi \, dx - \int_\Omega P_0 \cdot V_0 \phi \, dx$$
$$+ \int_\Omega (P_\varepsilon \cdot V_0)\phi \, dx + \int_\Omega P_0 \cdot V_\varepsilon \phi \, dx \tag{5.7}$$

We can pass to the limit in the last two terms on the right side of (5.7) by virtue of (5.4):

$$\int_\Omega (P_\varepsilon \cdot V_0)\phi \, dx \to \int_\Omega P_0 V_0 \phi \, dx \quad \text{and} \quad \int_\Omega (P_0 \cdot V_\varepsilon)\phi \, dx \to \int_\Omega P_0 V_0 \phi \, dx.$$

Then the right-hand side of (5.7) becomes

$$\int_\Omega (P_\varepsilon - P_0)(V_\varepsilon - V_0)\phi \, dx + \int_\Omega P_0 \cdot V_0 \phi \, dx + o(1), \text{ as } \varepsilon \to 0.$$

Thus, (5.6) holds if and only if $(P_\varepsilon - P_0)(V_\varepsilon - V_0) \rightharpoonup 0$ in $\mathscr{D}'(\Omega)$ as $\varepsilon \to 0$, and it is sufficient to consider the case where $P_0 = 0$, $V_0 = 0$.

Step 2: Showing That $V_\varepsilon = \nabla u_\varepsilon$ with $u_\varepsilon \to 0$ Strongly in $L^2(\Omega)$. We first observe that $V_\varepsilon = \nabla u_\varepsilon$ for some function u_ε. Indeed, since curl $V_\varepsilon = 0$, then V_ε is an irrotational field and is necessarily *potential*. That is it can be written as the gradient of a scalar function provided that the domain Ω is simply connected, otherwise, one can repeat the following arguments locally.

Since u_ε is defined up to a constant, we impose the condition that

$$\int_\Omega u_\varepsilon \, dx = 0. \tag{5.8}$$

From (5.4) we have

$$\nabla u_\varepsilon \rightharpoonup V_0 = 0 \text{ weakly in } L^2(\Omega). \tag{5.9}$$

Then by virtue of the Poincaré inequality (see Theorem 1.5) we have $\|u_\varepsilon\|_{H^1(\Omega)} \leq C$. Due to the compactness of the embedding of $H^1(\Omega)$ in $L^2(\Omega)$ this yields the strong convergence $u_\varepsilon \to 0$ in $L^2(\Omega)$.

Step 3: Establishing Convergence in $\mathcal{D}'(\Omega)$. For any scalar test function $\phi \in C_0^\infty(\Omega)$, we have

$$\int_\Omega P_\varepsilon V_\varepsilon \phi \, dx = \int_\Omega P_\varepsilon \nabla u_\varepsilon \phi \, dx = \int_\Omega P_\varepsilon \nabla(u_\varepsilon \phi) \, dx - \int_\Omega u_\varepsilon P_\varepsilon \nabla \phi \, dx$$

$$= -\int_\Omega \operatorname{div} P_\varepsilon \, \phi u_\varepsilon dx - (P_\varepsilon, u_\varepsilon \nabla \phi)_{L^2}. \tag{5.10}$$

Since $u_\varepsilon \rightharpoonup 0$ weakly in $H^1(\Omega)$ and $u_\varepsilon \to 0$ strongly in $L^2(\Omega)$, then $\phi u_\varepsilon \rightharpoonup 0$ weakly in $H^1(\Omega)$ and $u_\varepsilon \nabla \phi \to 0$ strongly in $(L^2(\Omega))^n$. Thus, using (5.4) and (5.5), we can pass to the limit in (5.10) concluding the proof of the Div-Curl Lemma 5.1. $\qquad \square$

5.2 Proof of Theorem 2.1 via Oscillating Test Functions and Div-Curl Lemma

We now prove the homogenization Theorem 2.1 using the Div-Curl Lemma 5.1. Recall the weak formulation of (2.5)–(2.6):

$$\int_\Omega \sigma\left(\frac{x}{\varepsilon}\right) \nabla u_\varepsilon \cdot \nabla \phi \, dx = \int g\phi \, dx, \quad \forall \phi \in H_0^1(\Omega). \tag{5.11}$$

The main difficulty in the proof is that $\sigma_\varepsilon \rightharpoonup \langle\sigma\rangle$ and $\nabla u_\varepsilon \rightharpoonup \nabla u_0$ weakly in $L^2(\Omega)$ (or $L^2(\Omega)^n$), but their product does not necessarily converge (weakly in $(L^2(\Omega))^n$) to the product of weak limits which is why one cannot directly pass to the limit in (5.11). To resolve this difficulty one can choose special test functions $\phi = \phi_\varepsilon \in H_0^1(\Omega)$ which depend on ε in such a way that we can apply the Div-Curl Lemma. Recall the following limits, which were established in ▶ Section 2 (see (2.8)):

$$u_\varepsilon \rightharpoonup u_0 \text{ weakly in } H^1(\Omega) \tag{5.12}$$

$$\nabla u_\varepsilon \rightharpoonup \nabla u_0 \text{ weakly in } L^2(\Omega) \tag{5.13}$$

$$u_\varepsilon \to u_0 \text{ strongly in } L^2(\Omega). \tag{5.14}$$

Given an arbitrary test function $\phi \in C_0^\infty(\Omega)$ (a dense subset of $H_0^1(\Omega)$), we construct a special set of oscillating test functions, ϕ_ε such that the following conditions hold:

(H1) $\phi_\varepsilon \rightharpoonup \phi$ weakly in $L^2(\Omega)$;

(H2) $\mathrm{div}\left(\sigma\left(\frac{x}{\varepsilon}\right)\nabla\phi_\varepsilon\right) \to \mathrm{div}\left(\hat{\sigma}\nabla\phi\right)$ strongly in $H^{-1}(\Omega)$;

(H3) $\sigma\left(\frac{x}{\varepsilon}\right)\nabla\phi_\varepsilon \rightharpoonup \hat{\sigma}\nabla\phi$ weakly in $L^2(\Omega)$.

Step 1: Passing to the Limit in (5.11) Under the Assumption That There Exists a Family of Test Functions Satisfying Properties (H1)–(H3). Set

$$P_\varepsilon := \sigma\left(\frac{x}{\varepsilon}\right)\nabla\phi_\varepsilon, \quad P_0 := \hat{\sigma}\nabla\phi.$$

Note that (H3) implies that $P_\varepsilon \rightharpoonup P_0$ weakly in $L^2(\Omega)$, and that (H2) implies that $\mathrm{div}\, P_\varepsilon \to \mathrm{div}\, P_0$ strongly in $H^{-1}(\Omega)$. Set $V_\varepsilon := \nabla u_\varepsilon$ and observe that the curl $V_\varepsilon = \mathrm{curl}\,\nabla u_\varepsilon = 0$. All of the hypotheses of the Div-Curl Lemma hold and thus we can pass to the limit in the product of weakly convergent sequences in the left-hand side of (5.11)

$$\int_\Omega \sigma\left(\frac{x}{\varepsilon}\right)\nabla u_\varepsilon \nabla\phi_\varepsilon dx \to \int_\Omega \hat{\sigma}\nabla u_0 \nabla\phi. \tag{5.15}$$

For the right-hand side of (5.11), use (H1) to pass to the limit, $\int g\phi_\varepsilon\, dx \to \int g\phi\, dx$. Thus,

$$\int_\Omega \hat{\sigma}\nabla u_0 \nabla\phi\, dx = \int_\Omega g\phi\, dx, \quad \forall \phi \in C_0^\infty(\Omega). \tag{5.16}$$

This holds for all test functions $\phi \in C_0^\infty(\Omega)$, and by density of $C_0^\infty(\Omega)$ in $H_0^1(\Omega)$, (5.16) holds for every $\phi \in H_0^1(\Omega)$. Thus, Theorem 2.1 (homogenization limit) is proved provided that we prove existence of functions ϕ_ε with properties (H1)-(H3).

Step 2: Construction of Oscillating Test Functions ϕ_ε. Given $\phi \in C_0^\infty(\Omega)$, set

$$\phi_\varepsilon(x) := \phi(x) + \varepsilon \sum_{i=1}^n \frac{\partial\phi}{\partial x_i}(x)\chi_i\left(\frac{x}{\varepsilon}\right) \tag{5.17}$$

where χ_i is a solution to the cell problem (2.14)–(2.15). Note that the functions ϕ_ε mimic a Taylor expansion since χ_i are coordinate-like functions. Condition (H1) follows immediately from the form of (5.17). Indeed, for all $\psi \in L^2(\Omega)$, the Cauchy-Schwarz inequality yields

$$\varepsilon\sum_{i=1}^n \int \frac{\partial\phi}{\partial x_i}\chi_i\left(\frac{x}{\varepsilon}\right)\psi \le \varepsilon\sum_{i=1}^n \|\phi\|_{C^1(\Omega)}\|\chi_i(x/\varepsilon)\|_{L^2(\Omega)}\|\psi\|_{L^2(\Omega)} \to 0, \tag{5.18}$$

since $\chi_i \in H_{per}^1(\Pi)$.

To prove (H3) observe that

$$\sigma\left(\frac{x}{\varepsilon}\right)\nabla\phi_\varepsilon = \sigma\left(\frac{x}{\varepsilon}\right)\left[\nabla_x\phi(x) + \sum_{i=1}^{n}\frac{\partial\phi}{\partial x_i}\nabla_y\chi_i\left(\frac{x}{\varepsilon}\right)\right] + \varepsilon\sum_{i=1}^{n}\sigma\left(\frac{x}{\varepsilon}\right)\nabla\frac{\partial\phi}{\partial x_i}\chi_i\left(\frac{x}{\varepsilon}\right),$$

(5.19)

where the L^2 norm of the last term is of order ε. Take the weak L^2 limit in the right-hand side of (5.19) using the Averaging Lemma (we assume that $|\Pi| = 1$ for simplicity):

$$\sigma\left(\frac{x}{\varepsilon}\right)\nabla\phi_\varepsilon \rightharpoonup \int_\Pi \sigma(y)[I + \nabla_y\chi(y)]dy \cdot \nabla\phi(x) = \hat\sigma\nabla\phi,$$

(5.20)

where the last equality in (5.20) follows from the definition of effective conductivity (4.20). Note that the first term in the right-hand side of (5.19) depends not only the "fast" variable $y = x/\varepsilon$ but also on the "slow" variable x. However one still can apply the Averaging Lemma using the fact that each term has the form of product of a smooth function depending on x only and periodic function depending on $\frac{x}{\varepsilon}$. Indeed, for example, considering the first term in (5.19) we have by Averaging Lemma

$$\int_\Omega \sigma\left(\frac{x}{\varepsilon}\right)\nabla_x\phi(x)\psi(x)\,dx \rightarrow \int_\Omega \left[\int_\Pi \sigma(y)\,dy\right]\nabla_x\phi(x)\psi(x)\,dx,$$

for arbitrary function $\psi \in L^2(\Omega)$, therefore $\sigma\left(\frac{x}{\varepsilon}\right)\nabla_x\phi(x) \rightharpoonup \hat\sigma\nabla_x\phi$. Thus we have proved (H3).

It remains to prove the key property (H2). To this end we compute

$$\text{div}\left[\sigma\left(\frac{x}{\varepsilon}\right)\nabla\phi_\varepsilon\right] = \frac{1}{\varepsilon}\sum_{i,j}\frac{\partial\phi}{\partial x_i}(x)\frac{\partial}{\partial y_j}\left[\sigma_{ij}(y)\left(\delta_{ij} + \frac{\partial\chi_i}{\partial y_j}(y)\right)\right]\Big|_{y=\frac{x}{\varepsilon}}$$

$$+ \sigma\left(\frac{x}{\varepsilon}\right)\Delta\phi(x) + \sigma\left(\frac{x}{\varepsilon}\right)\sum_{i,j}\frac{\partial\phi}{\partial x_i\partial x_j}\frac{\partial\chi_i}{\partial y_j}\left(\frac{x}{\varepsilon}\right)$$

$$+ \varepsilon\text{div}\left[\sigma\left(\frac{x}{\varepsilon}\right)\sum_i\left(\nabla\frac{\partial\phi}{\partial x_i}\right)(x)\chi_i\left(\frac{x}{\varepsilon}\right)\right] =: I_{-1}^{(\varepsilon)} + I_0^{(\varepsilon)} + I_1^{(\varepsilon)}.$$

The first term $I_{-1}^{(\varepsilon)}$ actually zero since functions χ_i are solutions of the cell problem. The second term $I_0^{(\varepsilon)}$ converges weakly in $L^2(\Omega)$ to

$$I_0^{(\varepsilon)} \rightharpoonup \int_\Pi \left[\sigma(y)\Delta\phi(x) + \sigma(y)\sum\frac{\partial\phi}{\partial x_i\partial x_j}\frac{\partial\chi_i}{\partial y_j}(y)\right]dy$$

(5.21)

$$= \sum_i\frac{\partial}{\partial x_i}\int_\Pi[\sigma(y)\frac{\partial\phi}{\partial x_i}(x) + \sigma(y)\sum_j\frac{\partial\phi}{\partial x_j}\frac{\partial\chi_i}{\partial y_j}(y)]dy$$

$$= \sum_i\sum_j\frac{\partial}{\partial x_i}\frac{\partial\phi}{\partial x_i}e_j\int_\Pi \sigma(y)\left[e_i + \nabla_y\chi_i\right]dy = \text{div}\left[\hat\sigma\nabla\phi\right].$$

As above, (5.21) can be proved by applying the Averaging Lemma. Thus $I_0^{(\varepsilon)}$ converges to div $\left[\hat{\sigma}\nabla\phi\right]$ strongly in $H^{-1}(\Omega)$. Indeed, weak convergence in $L^2(\Omega)$ implies boundedness in $L^2(\Omega)$ which in turn by the compactness of the embedding $L^2(\Omega) \subset H^{-1}(\Omega)$ implies strong convergence in $H^{-1}(\Omega)$. Finally, $I_1^{(\varepsilon)}$ converges to 0 strongly in $H^{-1}(\Omega)$. Indeed, $I_1^{(\varepsilon)}$ has the form $I_1^{(\varepsilon)} = \varepsilon \operatorname{div} F_\varepsilon$ with $F_\varepsilon := \sigma\left(\frac{x}{\varepsilon}\right) \sum \left(\nabla \frac{\partial\phi}{\partial x_i}\right)(x)\chi_i\left(\frac{x}{\varepsilon}\right)$. Therefore

$$\left|\langle \operatorname{div} F_\varepsilon, \varphi\rangle_{H^{-1},H_0^1}\right| = \left|\langle F_\varepsilon, \nabla\varphi\rangle_{L^2,L^2}\right| \leq \|F_\varepsilon\|_{L^2}\|\varphi\|_{H_0^1} \leq C\|\varphi\|_{H_0^1},$$

i.e., $\|I_1^{(\varepsilon)}\|_{H^{-1}} = O(\varepsilon)$. Thus conditions (H1), (H2), and (H3) are satisfied, and the proof of Theorem 2.1 (the homogenization limit) is complete. $\square$

Two-Scale Convergence

© Springer Nature Switzerland AG 2018
L. Berlyand, V. Rybalko, *Getting Acquainted with Homogenization and Multiscale*,
Compact Textbooks in Mathematics,
https://doi.org/10.1007/978-3-030-01777-4_6

Two-scale convergence is a version of weak convergence that is specifically suited for multiscale problems with periodic coefficients. For motivation, recall from Example 1.1 that a sequence of rapidly oscillating functions typically does not have pointwise convergence at any point. As such we turned to weak convergence as a way to find the limits of these functions. Even though the Averaging Lemma allows us to pass to the weak limit easily, it loses the structure of fine scale oscillations and only recovers the average value of the function. Ideally, we want to incorporate more information about the limiting function as we pass to the limit which is what the two-scale convergence approach is suited for. In fact G. Nguetseng invented this approach as a beautiful shortcut when working on problems for Biot's equations in poroelasticity models [80]. For such models the direct justification via asymptotic expansions (see ▶ Section 4) becomes immensely technical, and two-scale convergence was his way around it. The two-scale convergence method was further developed by G. Allaire [3] to cover a variety of linear and even some nonlinear periodic homogenization problems.

Throughout this section we assume for convenience that $|\Pi| = 1$.

Definition 6.1 (Two-Scale Convergence)

Let $\{u_\varepsilon\}$ be a sequence of functions from $L^2(\Omega)$. We say that u_ε two-scale converges to a limit $u_0(x, y) \in L^2(\Omega \times \Pi)$, and write

$$u_\varepsilon(x) \overset{2}{\rightharpoonup} u_0(x, y), \tag{6.1}$$

if $\{u_\varepsilon\}$ is bounded in $L^2(\Omega)$, and for any function $\psi(x, y) \in D(\Omega, C^\infty_{per}(\Pi))$ the following equality holds:

$$\lim_{\varepsilon \to 0} \int_\Omega u_\varepsilon(x)\psi(x, \frac{x}{\varepsilon})dx = \int_\Omega \int_\Pi u_0(x, y)\psi(x, y)dydx. \tag{6.2}$$

(continued)

Definition 6.1 (continued)

Here, $D(\Omega, C^\infty_{per}(\Pi))$ is the space of infinitely differentiable functions with compact support in Ω with respect to the first argument and taking values in $C^\infty_{per}(\Pi)$ with respect to the second argument.

ⓘ Remark 6.1 While some authors do not impose the requirement for the sequence $\{u_\varepsilon\}$ to be bounded in $L^2(\Omega)$, i.e.,

$$\|u_\varepsilon(x)\|_{L^2} \le C, \tag{6.3}$$

in the definition of two-scale convergence, it is more convenient to have (6.3) in the definition, e.g., for a comparison of the two-scale convergence with weak and strong convergences in $L^2(\Omega)$. Note that if one widens the class of test functions by adding to this class all the functions $\psi(x, y) = \psi(x) \in L^2(\Omega)$ (cf. definition of weak convergence, where test functions are elements of $L^2(\Omega)$), the uniform boundedness would not be required as part of the definition but can be established as a property of a two-scale converging sequence. However, in practice it is much more convenient to take smooth $\psi(x, y)$ as in Definition 6.1, e.g., because we often need to differentiate them. Further, uniform boundedness of u_ε is not an overly restrictive assumption and can be proven via a priori estimates in a wide class of multiscale problems.

Exercise 6.1

Show that if (6.3) holds, then the following definition is equivalent to (6.2): $\forall \phi \in C^\infty_0(\Omega)$, $\Phi \in C^\infty_{per}(\Pi)$

$$\lim_{\varepsilon \to 0} \int_\Omega u_\varepsilon(x)\phi(x)\Phi\left(\frac{x}{\varepsilon}\right) dx = \int_{\Omega \times \Pi} u_0(x, y)\phi(x)\Phi(y)\, dxdy. \tag{6.4}$$

■

The following theorem provides a second, practical reason for the condition of boundedness in Definition 6.1:

Theorem 6.1 (The First Compactness Theorem)
If a sequence $\{u_\varepsilon\}$ satisfies (6.3), then it contains a subsequence that two-scale converges to some $u_0(x, y) \in L^2(\Omega \times \Pi)$.

Proof

Consider the functional

$$\ell_\varepsilon(\psi) = \int_\Omega u_\varepsilon(x)\psi\left(x, \frac{x}{\varepsilon}\right) dx, \tag{6.5}$$

where $\psi(x, y)$ is a finite linear combination of functions $\phi(x)\Phi(y)$ with $\phi \in C_0^\infty(\Omega)$ and $\Phi \in C_{per}^\infty(\Pi)$. We can choose a countable subset of such functions which is dense in $L^2(\Omega \times \Pi)$. This subset will be referred to as the set of admissible test functions. Next, for any $\psi(x, y) = \sum_{\text{finite}} \lambda_k \phi_k(x)\Phi_k(y)$ we have

$$\int_\Omega \psi^2\left(x, \frac{x}{\varepsilon}\right) dx = \sum_{k,k'} \lambda_k \lambda_{k'} \int_\Omega \phi_k(x)\Phi_k\left(\frac{x}{\varepsilon}\right)\phi_{k'}(x)\Phi_{k'}\left(\frac{x}{\varepsilon}\right) dx$$

$$\overset{\substack{\text{by Averaging} \\ \text{Lemma}}}{\underset{\varepsilon \to 0}{\longrightarrow}} \sum_{k,k'} \lambda_k \lambda_{k'} \int_\Omega \phi_k(x)\phi_{k'}(x) \int_\Pi \Phi_k(y)\Phi_{k'}(y)\, dy\, dx$$

$$= \int_{\Omega \times \Pi} \psi^2(x, y)\, dx\, dy. \tag{6.6}$$

For every admissible function ψ we have

$$\limsup_{\varepsilon \to 0} \ell_\varepsilon(\psi) \overset{\text{Cauchy-Schwarz}}{\leq} \limsup_{\varepsilon \to 0} \|u_\varepsilon\|_{L^2} \left(\int_\Omega \psi^2\left(x, \frac{x}{\varepsilon}\right)\right)^{1/2}$$

$$\overset{\text{by (6.6)\&(6.3)}}{\leq} C\left(\int_{\Omega \times \Pi} \psi^2(x, y)\, dx\, dy\right)^{1/2}. \tag{6.7}$$

Since the space of admissible functions is countable, we can extract a subsequence $\varepsilon' \to 0$ such that $\ell_{\varepsilon'}(\psi)$ converges to a limit for all admissible test functions ψ. In this limit we obtain a linear functional $\ell(\psi)$ and by (6.7) this functional $\ell(\psi)$ can be continued to a bounded linear functional in the entire space $L^2(\Omega \times \Pi)$. By the Riesz representation theorem $\exists u_0 \in L^2(\Omega \times \Pi)$ such that

$$\ell(\psi) = \int_{\Omega \times \Pi} u_0(x, y)\psi(x, y)\, dx\, dy.$$

Thus

$$\int_\Omega u_{\varepsilon'}(x)\phi(x)\Phi\left(\frac{x}{\varepsilon'}\right) dx \to \int_{\Omega \times \Pi} u_0(x, y)\phi(x)\Phi(y)\, dx\, dy,$$

and using Exercise 6.1 we obtain the desired result. □

It is natural to ask the question "how does two-scale convergence relate to strong and weak L^2 convergence?" As shown below, two-scale convergence is somewhere in between strong and weak L^2 convergence. Before establishing these results we state the so-called generalized Averaging Lemma, whose proof is analogous to the Averaging Lemma and is left to the reader:

> ⓘ **Lemma 6.1 (Generalized Averaging Lemma)** *Assume that $f(x, y)$ is Π-periodic in y, and $f \in C(\Omega; C_{per}(\Pi))$. Then,*

$$\lim_{\varepsilon \to 0} \int_{\Omega} f\left(x, \frac{x}{\varepsilon}\right) g(x)\, dx = \int_{\Omega} \left(\int_{\Pi} f(x, y)\, dy\right) g(x)\, dx, \ \forall g \in L^2(\Omega).$$

Proposition 6.1 *Any sequence $\{u_\varepsilon\}$ that converges strongly in $L^2(\Omega)$ to $u_0(x)$, also two-scale converges to $u_0(x)$.*

Proof

We must show that

$$\int_{\Omega} u_\varepsilon(x)\psi\left(x, \frac{x}{\varepsilon}\right) dx - \int_{\Omega} \int_{\Pi} u_0(x)\psi(x, y)\, dydx \to 0 \tag{6.8}$$

as $\varepsilon \to 0$. We rewrite the left-hand side of (6.8) as

$$\int_{\Omega} (u_\varepsilon(x) - u_0(x))\psi\left(x, \frac{x}{\varepsilon}\right) dx + \left(\int_{\Omega} u_0(x) \left[\psi\left(x, \frac{x}{\varepsilon}\right) - \int_{\Pi} \psi(x, y)\, dy\right] dx\right)$$

$$=: I_1 + I_2. \tag{6.9}$$

Notice that

$$I_1 \leq C \int |u_\varepsilon(x) - u_0(x)|\, dx \leq C|\Omega|^{\frac{1}{2}} \|u_\varepsilon - u_0\|_{L^2} \to 0$$

by strong L^2 convergence. To complete the proof it remains to apply the generalized Averaging Lemma 6.1 with u_0 as a test function to $\psi\left(x, \frac{x}{\varepsilon}\right)$ to show that $I_2 \to 0$. □

To summarize, if the strong L^2 limit exists, then the two-scale limit also exists and the limits agree. In contrast, if the two-scale limit exists, then a weak L^2 limit also exists but these limits may be different. Namely, the weak L^2 limit can be obtained by averaging the two-scale limit in the y variable over its period, as the following example shows.

Example 6.1

Let $u_\varepsilon = \sin\left(\frac{x}{\varepsilon}\right)$, $\ x \in [0, 2\pi]$. Since $\Pi = [0, 2\pi]$ is a periodic cell, and u_ε is bounded, then we can apply the generalized Averaging Lemma to deduce

$$\int_0^{2\pi} \sin\left(\frac{x}{\varepsilon}\right) \phi(x) \Phi\left(\frac{x}{\varepsilon}\right) \to \int_{\Omega} \int_{\Pi} \sin(y)\phi(x)\Phi(y).$$

By definition (6.4) of two-scale convergence we deduce $\sin\left(\frac{x}{\varepsilon}\right) \overset{2}{\rightharpoonup} \sin(y)$. However, considering the weak limit, we can apply the (regular) Averaging Lemma to see that

$$\sin\left(\frac{x}{\varepsilon}\right) \rightharpoonup \langle\sin(y)\rangle = 0 \text{ weakly in } L^2(\Omega).$$

∎

Conversely, can we have a weakly convergent sequence that does not two-scale converge? The following example answers this question.

Example 6.2

Let

$$u_n = (-1)^n \sin(nx), \quad \varepsilon = \frac{1}{n}.$$

In the weak sense, we know that u_n converges to

$$\int_0^{2\pi} (-1)^n \sin\left(\frac{x}{1/n}\right) dx = \langle \sin(y) \rangle = 0.$$

If $n = 2k, k \in \mathbb{N}$, then by the generalized Averaging Lemma, we have that $u_{2k} \xrightarrow{2} \sin(y)$. However, when $n = 2k + 1$, then we have $u_{2k+1} \xrightarrow{2} -\sin(y)$. Therefore, a two-scale limit for u_n does not exist. ∎

The two-scale convergence method is suited for analysis of asymptotic expansions (i.e., ansatz) of the form

$$u_0\left(x, \frac{x}{\varepsilon}\right) + \varepsilon u_1\left(x, \frac{x}{\varepsilon}\right) + \varepsilon^2 u_2\left(x, \frac{x}{\varepsilon}\right) + \dots$$

Note that in the homogenized solution $u_0 = u_0(x)$ to the case study conductivity problem (2.5)–(2.6) does not depend on y, but this is not usually the case. In our previous example we saw that the weak limit preserves only the average value of oscillating functions; however, the two-scale limit captures oscillations, for instance, two-scale limits of fluxes do contain fast oscillations with period ε. Some properties of two-scale convergence listed in the following proposition; interested readers can find the proof of this proposition, e.g., in [3].

Proposition 6.2 *Let u_ε be a sequence of $L^2(\Omega)$ functions that two-scale converges to $u_0 \in L^2(\Omega \times \Pi)$, then*

$$u_\varepsilon(x) \rightharpoonup u(x) = \frac{1}{|\Pi|} \int_\Pi u_0(x, y) dy \tag{6.10}$$

and $\liminf_{\varepsilon \to 0} \|u_\varepsilon\|_{L^2(\Omega)} \geq \|u_0\|^2_{L^2(\Omega \times \Pi)} \geq \|u\|^2_{L^2(\Omega)}.$ \tag{6.11}

If, in addition, $u_0(x, y)$ is smooth, and

$$\lim_{\varepsilon \to 0} \|u_\varepsilon\|_{L^2(\Omega)} = \|u_0\|_{L^2(\Omega \times \Pi)} \tag{6.12}$$

then $u_\varepsilon(x) - u_0\left(x, \frac{x}{\varepsilon}\right) \to 0$ *strongly in $L^2(\Omega)$.* \tag{6.13}

ⓘ Remark 6.2 Let us summarize the relations between weak L^2, strong L^2, and two-scale convergences:

— strong L^2 convergence implies two-scale convergence (Proposition 6.1);
— two-scale convergence implies weak L^2 convergence (Proposition 6.2).

Now suppose that the L^2 boundedness condition (6.3) is replaced by H^1 boundedness,

$$\|u_\varepsilon\|_{H^1(\Omega)} \le C. \tag{6.14}$$

Then we can establish the following compactness result for u_ε and ∇u_ε.

Theorem 6.2 (Second Compactness Theorem)
If the condition (6.14) holds, then $\{u_\varepsilon\}$ contains a subsequence $\{u_{\varepsilon_k}\}$ (by abuse of notation also denoted by u_ε) such that

$$u_\varepsilon(x) \overset{2}{\rightharpoonup} u_0(x) \tag{6.15}$$

$$\nabla_x u_\varepsilon(x) \overset{2}{\rightharpoonup} \nabla_x u_0(x) + \nabla_y u_1(x, y), \tag{6.16}$$
where $u_1(x, y) \in L^2(\Omega, H^1_{per}(\Pi))$.

Remark: In the first compactness theorem, we assume only L^2 boundedness and cannot say if the limit contains fine scales. The second compactness theorem guarantees that due to H^1 boundedness, the limit u_0 has only the coarse scale.

Proof
It follows from the assumption (6.14) and the first compactness theorem (Theorem 6.1) that there exists $u_0(x, y) \in L^2(\Omega \times \Pi)$ and $v(x, y) \in \left(L^2(\Omega \times \Pi)\right)^n$ such that, up to extracting a subsequence,

$$u_\varepsilon \overset{2}{\rightharpoonup} u_0(x, y) \tag{6.17}$$

$$\nabla u_\varepsilon \overset{2}{\rightharpoonup} v(x, y). \tag{6.18}$$

In particular, for any $\varphi(x) \in C_0^\infty(\Omega)$, $\phi(y) \in C_{per}^\infty(\Pi)$, we have

$$\int_\Omega \nabla u_\varepsilon(x, y)\varphi(x) \cdot \phi\left(\frac{x}{\varepsilon}\right) dx \to \int_{\Omega \times \Pi} v(x, y)\varphi(x) \cdot \phi(y)\, dy dx. \tag{6.19}$$

Note that in this proof the functions ϕ are vector-valued. Multiply the left-hand side of (6.19) by ε and integrate by parts to get

$$\varepsilon \int_\Omega \nabla u_\varepsilon \left(x, \frac{x}{\varepsilon}\right) \varphi(x) \cdot \phi \left(\frac{x}{\varepsilon}\right) dx = - \int_\Omega u_\varepsilon(x)\varphi(x) \operatorname{div} \phi \left(\frac{x}{\varepsilon}\right) dx$$

$$- \varepsilon \int_\Omega u_\varepsilon(x)\nabla_x\varphi(x) \cdot \phi \left(\frac{x}{\varepsilon}\right) dx. \tag{6.20}$$

Then take the limit as $\varepsilon \to 0$ in (6.20) and use (6.17) to obtain

$$- \int_\Omega \int_\Pi u_0(x, y)\varphi(x)\operatorname{div}\phi(y)\, dy dx = 0. \tag{6.21}$$

Rearranging (6.21) with the help of Fubini's theorem yields

$$- \int_\Omega \left(\int_\Pi u_0(x, y)\operatorname{div}\phi(y)dy \right) \varphi(x)dx = 0. \tag{6.22}$$

By the du Bois-Reymond Lemma 1.2 we deduce that

$$\int_\Pi u_0(x, y)\operatorname{div}\phi(y)dy = 0. \tag{6.23}$$

Integrating by parts in (6.23) and another application of the du Bois-Reymond Lemma proves

$$\nabla_y u_0(x, y) = 0 \text{ and therefore } u_0(x, y) = u_0(x). \tag{6.24}$$

Similarly to the above proofs of two-scale convergence properties, we now choose appropriate test functions. Namely, in (6.19) we consider ϕ from the class of test functions such that $\operatorname{div}\phi(y) = 0$. Then, integrate by parts and use (6.17):

$$- \int_\Omega \int_\Pi v(x, y)\varphi(x) \cdot \phi(y)\, dy dx \overset{\text{by (6.19)}}{=} \lim_{\varepsilon \to 0} \int_\Omega u_\varepsilon \left(x, \frac{x}{\varepsilon}\right) \nabla_x\varphi(x) \cdot \phi \left(\frac{x}{\varepsilon}\right) dx$$

$$= \int_\Omega \int_\Pi u_0(x)\nabla_x\varphi(x) \cdot \phi(y)\, dy dx \tag{6.25}$$

Now integrate by parts in x in the right-hand side of (6.25). We obtain

$$\int_\Omega \int_\Pi [v(x, y) - \nabla u_0(x)]\varphi(x) \cdot \phi(y)\, dy dx = 0, \tag{6.26}$$

for all $\varphi \in C_0^\infty(\Omega)$ and $\phi \in C_{per}^\infty(\Pi)$ such that $\operatorname{div}\phi = 0$. Now recall that the orthogonal complement of gradients of functions from $H_{per}^1(\Pi)$ in $L^2(\Omega)$ is exactly the set of divergence free functions,

$$\{g \in H_{per}^1(\Pi): g = \nabla_y f, \ f \in L^2(\Pi)\}^\perp = \{\phi \in L^2(\Pi): \operatorname{div}\phi = 0\}. \tag{6.27}$$

Thus, if $\int_\Pi g\phi = 0$ for all divergence free functions $\phi \in C_{per}^\infty(\Pi)$, then by density $\int_\Pi g\phi = 0$ for all divergence free $\phi \in L^2(\Pi)$, therefore $g = \nabla_y f$ for some $f \in L^2(\Pi)$. Since $\varphi(x)$ in (6.26) is an arbitrary from C_0^∞ we conclude that $v(x, y) - \nabla u_0(x) = \nabla_y u_1(x, y)$. □

6.1   The First Application of Two-Scale Convergence: Proof of Theorem 2.1

We now present another proof of the homogenization theorem 2.1 for the case study conductivity problem. In the proof based on compensated compactness (see ▶ Section 5) there were two main steps: the formal asymptotic expansion and an application of the Div-Curl Lemma. We will see that the proof of this theorem using two-scale convergence is a shorter, one step procedure and thus can be easier applied to vectorial problems such as fluids and elasticity.

Consider again the weak formulation of the case study problem:

$$\sum_{i,j=1}^{n} \int_{\Omega} \sigma_{ij}\left(\frac{x}{\varepsilon}\right) \frac{\partial u_\varepsilon}{\partial x_j} \frac{\partial v}{\partial x_i} dx = \int_{\Omega} g(x)v(x)dx, \qquad \forall v \in H_0^1(\Omega). \tag{6.28}$$

From the a priori bound (2.8) we have (up to a subsequence)

$$u_\varepsilon \overset{2}{\rightharpoonup} u_0(x) \tag{6.29}$$

$$\nabla_x u_\varepsilon \overset{2}{\rightharpoonup} \nabla_x u_0(x) + \nabla_y u_1(x, y). \tag{6.30}$$

Now choose $v := \varepsilon\varphi(x)\phi\left(\frac{x}{\varepsilon}\right)$ in (6.28), where $\varphi(x) \in C_0^\infty(\Omega)$, and $\phi(y) \in C_{per}^\infty(\Pi)$,

$$\sum_{i,j=1}^{n} \int_{\Omega} \sigma_{ij}\left(\frac{x}{\varepsilon}\right) \frac{\partial u_\varepsilon}{\partial x_j} \left(\varphi(x)\frac{\partial \phi}{\partial y_i}\left(\frac{x}{\varepsilon}\right) + \varepsilon\phi\left(\frac{x}{\varepsilon}\right)\frac{\partial \varphi}{\partial x_i}\right) dx = \varepsilon \int_{\Omega} g(x)\varphi(x)\phi\left(\frac{x}{\varepsilon}\right) dx. \tag{6.31}$$

Take the $\varepsilon \to 0$ limit in (6.31), to see that

$$\sum_{i,j=1}^{n} \int_{\Omega} \int_{\Pi} \sigma_{ij}(y) \left[\frac{\partial u_0(x)}{\partial x_j} + \frac{\partial u_1(x, y)}{\partial y_j} \right] \frac{\partial \phi}{\partial y_i}(y)\varphi(x)dydx = 0. \tag{6.32}$$

Here, we apply the definition of two-scale convergence considering $\varphi(x)$, $\sigma_{ij}(y) \partial_{y_j}\phi(y)$ as test functions with separated variables (as in Exercise 6.1). Note that these functions still can be used in the definition of two-scale convergence despite the fact that $\sigma_{ij}(y)$ do not belong to $C_{per}^\infty(\Pi)$ but rather to $L_{per}^\infty(\Pi)$ (see remarks in [3]). By virtue of du Bois-Reymond Lemma we conclude from (6.32) that for almost all $x \in \Omega$

$$\int_{\Pi} \sigma(y) \left(\nabla_x u_0(x) + \nabla_y u_1(x, y)\right) \cdot \nabla_y \phi(y) \, dy = 0 \,\, \forall \phi \in H_{per}^1(\Pi), \tag{6.33}$$

or

$$-\mathrm{div}_y \left(\sigma(y)(\nabla_x u_0(x) + \nabla_y u_1(x, y))\right) = 0 \text{ in } \Pi, \tag{6.34}$$

where $x \in \Omega$ is considered as a parameter and Π-periodicity of in y supplies (6.34) with boundary conditions for $u_1(x, y)$. In other words (6.33) is simply the weak formulation of (6.34) with periodic boundary conditions. Recall that the cell problem from Theorem 2.1 in PDE form (scalar equations for χ_i) reads

$$-\operatorname{div}\left[\sigma \nabla \chi_i + e_i\right] = 0, \qquad i = 1, \ldots, n.$$

It follows from (6.34) that $u_1(x, y)$ can be represented as

$$u_1(x, y) = \sum_{i=1}^{n} \partial_{x_i} u_0(x) \chi_i(y) + \text{ some function depending on } x \text{ only,}$$

so that

$$\nabla_y u_1(x, y) = \sum_{i=1}^{n} \partial_{x_i} u_0(x) \nabla_y \chi_i(y). \tag{6.35}$$

Now, take the limit in (6.28) as $\varepsilon \to 0$, with smooth $v(x)$ that do not depend on ε or y. Considering $\sigma_{ij}(y)\partial_{x_j} v(x)$ as test functions in the definition of two-scale convergence, we get

$$\int_{\Omega} \int_{\Pi} \sigma(y) \left(\nabla_x u_0(x) + \nabla_y u_1(x, y)\right) \cdot \nabla v(x) \, dxdy = \int_{\Omega} gv dx. \tag{6.36}$$

Substituting (6.35) into (6.36), we have

$$-\sum_{i,j=1}^{n} \int_{\Omega} \int_{\Pi} \left(\sigma_{ij}(y) + \sum_{k=1}^{n} \sigma_{ik}(y)\partial_{y_k}\chi_j(y)\right) \partial_{x_j} u_0 \partial_{x_i} v \, dx = \int_{\Omega} g(x)v(x)dx. \tag{6.37}$$

Now, recall the flux formula (4.27) for $\hat{\sigma}_{ij}$:

$$\hat{\sigma}_{ij} = \int \left[\sigma_{ij} + \sum_{k=1}^{n} \frac{\partial \chi_j}{\partial x_k} \sigma_{ik}(y)\right] dy$$

and use it in (6.37) to conclude

$$\int \hat{\sigma}_{ij} \frac{\partial u}{\partial x_j} \frac{\partial v}{\partial x_i} dx = \int_{\Omega} g(x)v(x)dx. \tag{6.38}$$

The latter relation is nothing but the weak formulation of the homogenized problem. □

Remark 6.3 The above proof becomes immediate, thanks to the second compactness theorem for two-scale convergence, Theorem 6.2 (cf. direct proof by asymptotic two-scale expansions [16, 62]). We next present an interesting application of two-scale convergence techniques to a high contrast stationary diffusion, see [3] for details. Direct proof by two-scale expansions for this problem would be prohibitively cumbersome.

6.2   Two-Scale Convergence in the Double-Porosity Model of Flow Through Fractured Reservoirs

We next apply two-scale convergence techniques to study a model of *double porosity*. This model was developed to study the flow of fluid within a naturally fractured reservoir e.g., petroleum reservoirs [5, 14]. The flow of fluids through fractured media is fundamentally different from the flow of fluids through unfractured media due to the very different porosities of the connected background (e.g., fractures) and the rock. ◻ Figure 6.1 is a schematic illustration of such a fractured rock.

◻ **Fig. 6.1** Fractured phase is depicted in white and solid rock phase (inclusions) is depicted in black.

In contrast to the case study conductivity problem (2.5)–(2.6), the solution of the homogenized double-porosity problem depends explicitly on both the micro- and macroscales. For simplicity we consider a time-independent analog of the time-dependent model proposed in [5, 58]. We also assume that the microstructure is periodic. These two simplifications capture the essence of the problem while allowing to avoid technicalities due to time dependence and non-periodicity.

Let $\Pi := (0, 1)^n$ be the periodicity cell, and assume that it splits into the subdomains Π_1 and Π_2, where Π_2 is compactly (with its closure) contained in the cell $(0, 1)^n$ and Π_1 is connected, see ◻ Figure 6.2. In other words our inclusions form a disperse structure, when individual inclusions are separated from each other (that is, for example, a connected system of channels is not acceptable). Now let $\chi_i(y)$ be the characteristic functions on the domains Π_i, i.e.,

$$\chi_i(y) = \begin{cases} 1 & y \in \Pi_i \\ 0 & y \notin \Pi_i, \end{cases}$$

periodically extended on the entire space $\mathbb{R}^n$. Consider subdomains $\Omega_\varepsilon^i = \{x \in \Omega : \chi_i\left(\frac{x}{\varepsilon}\right) = 1\}$, so that $\Omega = \Omega_\varepsilon^1 \cup \Omega_\varepsilon^2$, see ◻ Figure 6.2.

69

6

6.2 · Two-Scale Convergence in the Double-Porosity Model of Flow Through...

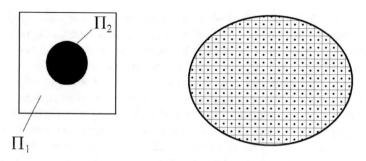

■ **Fig. 6.2** Mathematical model of double porosity: (left) periodicity cell and (right) domain with periodic microstructure.

We regard the connected domain Ω_ε^1 as representing the background medium, and the multiply connected domain Ω_ε^2 representing the inclusions. Define the diffusion coefficient μ_ε of the heterogeneous medium Ω by

$$\mu_\varepsilon(x) := \mu_1 \chi_1 \left(\frac{x}{\varepsilon}\right) + \varepsilon^2 \mu_2 \chi_2 \left(\frac{x}{\varepsilon}\right) \tag{6.39}$$

where μ_1 and μ_2 are positive constants. Let $f \in L^2(\Omega)$ be a given source term and consider the boundary value problem for a scalar u_ε:

$$-\text{div}\left[\mu_\varepsilon(x)\nabla u_\varepsilon(x)\right] + \alpha u_\varepsilon = f(x), \qquad x \in \Omega \tag{6.40}$$

$$u_\varepsilon = 0, \qquad x \in \partial\Omega. \tag{6.41}$$

For technical simplicity throughout the proof of the homogenization limit of this problem, we assume that $\alpha > 0$. The case where $\alpha = 0$ makes more sense physically, but is harder to work with technically. On the interface between Ω_ε^1 and Ω_ε^2 we have the following transmission conditions:

$$u_\varepsilon^+ = u_\varepsilon^-, \qquad \mu_\varepsilon^+ \frac{\partial u_\varepsilon^+}{\partial v} = \mu_\varepsilon^- \frac{\partial u_\varepsilon^-}{\partial v}, \tag{6.42}$$

which represents continuity of solution and the flux through the interface.

Exercise 6.2
Show that the boundary conditions (6.42) are natural boundary conditions in the minimization problem

$$\min_{v \in H_0^1(\Omega)} \int_\Omega \left(\frac{1}{2}\mu_\varepsilon|\nabla v|^2 + \frac{1}{2}\alpha v^2 - f v\right) dx. \tag{6.43}$$

Hint: Consider the problem on Ω_ε^1 and Ω_ε^2 separately. ∎

Boundary value problem (6.39)–(6.41) is a stationary version of well-known double-porosity problem describing fluid flow in fissured porous media. The unknown u_ε represents fluid pressure. We consider a reservoir Ω composed of two types of porous media: subdomain Ω_ε^1 represents fissured part of the rock with permeability $\mu_1 \sim 1$ and Ω_ε^2 represents a porous rock with low permeability ($\varepsilon^2\mu_2 \ll 1$). Another possible physical interpretation is thermal diffusion. High ratio between two permeability coefficients implies that we deal with heterogeneous media with high contrast properties. The specific scaling (the ratio between two permeabilities $\sim \varepsilon^2$) is chosen so that it allows for significant variations of u_ε on microscale ε in Ω_ε^2 so that the homogenized problem and its solution $u_0 = u_0(x, y)$ depends on both fine and coarse (macroscopic) scales. This stands in contrast with the homogenized case study problem (2.9)–(2.10) that does not contain the fine scale $y = x/\varepsilon$. Also, the leading 0-order term in the asymptotic expansion (4.4) for the case study problem does not depend on $y = x/\varepsilon$ which is not the case in the double-porosity problem.

We are now ready to begin deriving the homogenized problem. Since we employ two-scale convergence, it is natural that our first step is to establish a priori bounds.

Step 1: A Priori Bounds. Multiply (6.40) by u_ε and integrate by parts to get

$$\int_\Omega \mu_\varepsilon |\nabla u_\varepsilon|^2 dx + \alpha \int_\Omega u_\varepsilon^2 dx = \int_\Omega f u_\varepsilon dx. \tag{6.44}$$

Exercise 6.3

Derive from (6.44) the following a priori estimates for $\alpha > 0$:

$$\|u_\varepsilon\|_{L^2(\Omega)} \le C, \tag{6.45}$$

$$\|\nabla u_\varepsilon\|_{L^2(\Omega_\varepsilon^1)} \le C, \tag{6.46}$$

$$\|\nabla u_\varepsilon\|_{L^2(\Omega_\varepsilon^2)} \le \frac{C}{\varepsilon}. \tag{6.47}$$

Note that the last estimate (6.47) is not a uniform bound in ε, so the second compactness theorem does not apply. This is one of the special features of the problem distinguishing it from ▶ Section 6.1.   ∎

Since Ω_ε^1 is a connected subdomain in Ω, thanks to the bounds (6.45)–(6.46) there exists an extension $\hat{u}_\varepsilon \in H_0^1(\Omega)$ of u_ε from Ω_ε^1 onto Ω with $\hat{u}_\varepsilon = u_\varepsilon$ in Ω_ε^1 such that

$$\|\hat{u}_\varepsilon\|_{H^1(\Omega)} \le C, \tag{6.48}$$

with constant C independent of ε. Next we outline the extension construction; general techniques for such extensions can be found in [1]. We first describe this extension construction for a function $u(y) \in H^1(\Pi_1)$ in a rescaled unit cell Π depicted on ▢

Figure 6.2. Since Π_2 is compactly (with its closure) contained in Π (disperse structure consisting of isolated inclusions), we take $\hat{u}$ inside Π_2 to be the solution to the Laplace equation

$$\Delta_y \hat{u} = 0 \text{ in } \Pi_2 \tag{6.49}$$

$$\hat{u} = u \text{ on } \partial\Pi_2. \tag{6.50}$$

Then using the variational formulation of (6.49), Poincaré inequality, and the Trace Theorem 1.3 for a priori bounds on $\partial\Pi_2$, we see that the extension $\hat{u}$ satisfies

$$\|\nabla\hat{u}\|_{L^2(\Pi)} \le C\|\nabla u\|_{L^2(\Pi_1)}. \tag{6.51}$$

We construct such an extension $\hat{u}_\varepsilon$ for each periodicity cell (of size ε) in Ω to define $\hat{u}_\varepsilon$ in the entire domain Ω. Summing squares of L^2-norms of $\nabla\hat{u}_\varepsilon$ over all these cells and using (6.51) we obtain $\|\nabla\hat{u}_\varepsilon\|_{L^2(\Omega)} \le C\|u_\varepsilon\|_{H^1(\Omega_\varepsilon^1)}$. Then (6.46) implies (6.48). Thus, the desired extension is constructed.

It follows from Theorem 6.2 that there exist $u(x) \in H_0^1(\Omega), u_1(x, y) \in L^2(\Omega, H^1_{per}(\Pi))$ such that, up to a subsequence,

$$\hat{u}_\varepsilon \rightharpoonup u(x) \in H_0^1(\Omega), \tag{6.52}$$

$$\nabla\hat{u}_\varepsilon \overset{2}{\rightharpoonup} \nabla u(x) + \nabla_y u_1(x, y). \tag{6.53}$$

Since the sequence $\{u_\varepsilon\}$ is bounded in $L^2(\Omega)$ by (6.45), we also know that (possibly passing to a further subsequence)

$$u_\varepsilon \overset{2}{\rightharpoonup} u_0(x, y) \in L^2(\Omega \times \Pi). \tag{6.54}$$

Step 2: Relations Between Two-Scale Limits. Since we have no uniform bounds on ∇u_ε in $L^2(\Omega)$ but rather only on $\varepsilon\nabla u_\varepsilon$ from (6.46)–(6.47), we need the following proposition which relates $u_0(x, y)$ (from (6.54)) and the two-scale limit of $\varepsilon\nabla u_\varepsilon$.

Proposition 6.3 *Assume that $u_\varepsilon \overset{2}{\rightharpoonup} u_0(x, y)$ and that the sequence $\{\varepsilon\nabla u_\varepsilon\}$ is bounded in $(L^2(\Omega))^n$, then $u_0 \in L^2(\Omega; H^1_{per}(\Pi))$ and $\varepsilon\nabla u_\varepsilon \overset{2}{\rightharpoonup} \nabla_y u_0(x, y)$.*

Proof
Since the sequence $\{\varepsilon\nabla u_\varepsilon\}$ is bounded in $(L^2(\Omega))^n$, upon taking a subsequence we have

$$\varepsilon\nabla u_\varepsilon \overset{2}{\rightharpoonup} v(x, y) \in (L^2(\Omega \times \Pi))^n. \tag{6.55}$$

Let $\varphi \in C_0^\infty(\Omega), \phi \in (C_{per}^\infty(\Pi))^n$. Integrating by parts,

$$\int_\Omega \varepsilon \nabla u_\varepsilon \cdot \phi\left(\frac{x}{\varepsilon}\right) \varphi(x)\, dx = -\varepsilon \int_\Omega u_\varepsilon \nabla \varphi(x) \cdot \phi\left(\frac{x}{\varepsilon}\right) dx - \int_\Omega u_\varepsilon \operatorname{div}_y \phi\left(\frac{x}{\varepsilon}\right) \varphi(x)\, dx.$$

$$(6.56)$$

By the assumption $u_\varepsilon \overset{2}{\rightharpoonup} u_0(x, y)$, the first term on the right-hand side of (6.56) vanishes as $\varepsilon \to 0$. Thus, taking the limit $\varepsilon \to 0$ we obtain

$$\lim_{\varepsilon \to 0} \int_\Omega \varepsilon \nabla u_\varepsilon \cdot \phi\left(\frac{x}{\varepsilon}\right) \varphi(x)\, dx = -\int_\Omega \int_\Pi u_0(x, y)(\operatorname{div}_y \phi)(y)\, \varphi(x)\, dy dx$$

$$= \int_\Omega \int_\Pi \nabla_y u_0(x, y) \cdot \phi(y)\varphi(x)\, dy dx. \qquad (6.57)$$

Comparing (6.57) and (6.55) we see that $\nabla_y u_0(x, y) = v(x, y) \in \left(L^2(\Omega \times \Pi)\right)^n$. Moreover $u_0(x, \cdot) \in H^1_{per}(\Pi)$ for almost all $x \in \Omega$.   □

Applying Proposition 6.3 we immediately obtain that

$$\varepsilon \chi_2(x/\varepsilon)\nabla u_\varepsilon \overset{2}{\rightharpoonup} \chi_2(y)\nabla_y u_0(x, y). \qquad (6.58)$$

Also, since $\hat{u}_\varepsilon(x) \overset{2}{\rightharpoonup} u(x)$, we have $u_\varepsilon(x)\chi_1(x/\varepsilon) = \hat{u}_\varepsilon(x)\chi_1(x/\varepsilon) \overset{2}{\rightharpoonup} u(x)\chi_1(y)$. Thus

$$u_0(x, y) = u(x) \text{ in } \Pi_1, \qquad (6.59)$$

in particular, on the boundary $\partial \Pi_2$

$$u_0(x, y) = u(x) \text{ on } \partial \Pi_2. \qquad (6.60)$$

Step 3: Derivation of Homogenized Equations. Consider test functions $\varphi \in C^\infty_0(\Omega)$ and $\phi \in C^\infty_{per}(\Pi)$ such that $\phi(y) = 0$, when $y \in \Pi_1$. Using test function $\varphi(x)\phi(x/\varepsilon)$ in the weak formulation of (6.40) we obtain

$$\varepsilon \int_{\Omega^2_\varepsilon} \mu_2 \nabla u_\varepsilon \cdot (\nabla \phi)\left(\frac{x}{\varepsilon}\right) \varphi(x)\, dx + \alpha \int_{\Omega^2_\varepsilon} u_\varepsilon \phi\left(\frac{x}{\varepsilon}\right) \varphi(x)\, dx = \int_{\Omega^2_\varepsilon} f\phi\left(\frac{x}{\varepsilon}\right) \varphi(x)\, dx$$

$$- \varepsilon^2 \mu_2 \int_{\Omega^2_\varepsilon} \nabla u_\varepsilon \cdot (\nabla \varphi)(x)\phi\left(\frac{x}{\varepsilon}\right) dx. \qquad (6.61)$$

We can pass to the limit in this relation, using (6.58) and (6.54). The resulting relation

$$\mu_2 \int_{\Omega \times \Pi_2} \nabla_y u_0(x, y) \cdot \nabla_y \phi(y)\, \varphi(x)\, dy dx + \alpha \int_{\Omega \times \Pi_2} u_0(x, y)\phi(y)\, \varphi(x)\, dy dx$$

$$= \int_{\Omega \times \Pi_2} f\phi(y)\, \varphi(x)\, dy dx \qquad (6.62)$$

73

6

6.2 · Two-Scale Convergence in the Double-Porosity Model of Flow Through...

implies that for almost all $x \in \Omega$, $u_0(x, y)$ solves

$$-\mu_2 \Delta_y u_0(x, y) + \alpha u_0(x, y) = f(x), \quad y \in \Pi_2, \tag{6.63}$$

with boundary condition (6.60). Next consider $\phi \in C^\infty_{per}(\Pi)$ which is not necessarily 0 on Π_1. We take the test function $\psi(x) + \varepsilon \varphi(x) \phi(x/\varepsilon)$, with $\psi \in C^\infty_0(\Omega)$ in the weak formulation of (6.40) to obtain, after passing to the limit $\varepsilon \to 0$,

$$\mu_1 \int_{\Omega \times \Pi_1} (\nabla u(x) + \nabla_y u_1(x, y)) \cdot (\nabla \psi(x) + \nabla_y \phi(y) \varphi(x)) \, dy dx = \int_\Omega f \psi(x) \, dx$$

$$- \alpha \int_{\Omega \times \Pi} u_0(x, y) \psi(x) \, dy dx, \tag{6.64}$$

where we have used (6.47), (6.54) and the fact that $\chi_1(x/\varepsilon) \nabla u_\varepsilon \overset{2}{\rightharpoonup} \chi_1(y)(\nabla u(x) + \nabla_y u_1(x, y))$ which follows from (6.53). Now as in the proof of Theorem 2.1 in the previous section, we can show that $\nabla_y u_1(x, y)$ is the linear combination

$$\nabla_y u_1(x, y) = \sum_{i=1}^n \partial_{x_i} u \nabla w_i(y) \tag{6.65}$$

of solutions ∇w_i of problems

$$w_i \in H^1_{per}(\Pi_1), \quad \int_{\Pi_1} (\mathbf{e}_i + \nabla w_i(y)) \cdot \nabla \phi(y) \, dy = 0 \quad \forall \phi \in H^1_{per}(\Pi_1). \tag{6.66}$$

Here $H^1_{per}(\Pi_1)$ denotes the space of restrictions to Π_1 of functions from $H^1_{per}(\Pi)$. It remains to substitute (6.65) into (6.64) to obtain the homogenized equation for the background medium (high permeability phase):

$$-\mu_1 \mathrm{div} \, (\sigma^* \nabla u(x)) = f(x) - \alpha \int_\Pi u_0(x, y) \, dy. \tag{6.67}$$

Combining (6.67) with equations (6.63), (6.67), equality (6.59) and boundary conditions (6.60) and $u = 0$ on $\partial \Omega$ we arrive at closed system for the homogenized problem. The components of homogenized tensor σ^* in (6.67) are given by

$$\sigma^*_{ij} = \int_{\Pi_1} \mathbf{e}_i \cdot (\mathbf{e}_j + \nabla w_j(y)) \, dy.$$

The result is summarized as follows.

Theorem 6.3

The sequence u_ε of solutions of (6.40)–(6.41) two-scale converges to $u(x) + \chi_2(y)v(x, y)$, where (u, u) is the unique solution in $H_0^1(\Omega) \times L^2(\Omega \times H_{per}^1(\Pi_2))$ of the homogenized problem

$$-\mu_1 \mathrm{div}_x(\sigma^* \nabla_x u(x)) + \alpha u(x) = f(x) - \alpha \int_{\Pi_2} v(x, y) dy \quad \text{in } \Omega \tag{6.68}$$

$$-\mu_2 \Delta_{yy} v(x, y) + \alpha v(x, y) = f(x) - \alpha u(x) \text{ in } \Pi_2 \tag{6.69}$$

$$u = 0 \text{ on } \partial\Omega \quad \text{and} \quad v(x, y) = 0 \text{ on } \partial\Pi_2$$

where the entries of σ^* are given by

$$\sigma_{ij}^* = \int_{\Pi_1} \mathbf{e}_i \cdot (\mathbf{e}_j + \nabla w_j(y)) dy$$

and $w_i(y)$ are the solutions of the cell problems

$$-\mathrm{div}_y[\nabla_y w_i(y) + \mathbf{e}_i] = 0 \quad \text{in } \Pi_1$$

$$(\nabla_y w_i + \mathbf{e}_i) \cdot n = 0 \quad \text{on } \partial\Pi_2$$

$$w_i(y) \text{ is } \Pi - \text{periodic}$$

ⓘ Remark 6.4 (Simplification of the System (6.68)–(6.69)) Since the right-hand side of (6.69) is independent of y, one can write that $v(x, y) = [f(x) - \alpha u(x)]w(y)$, where $w(y)$ is the unique solution in $H_{per}^1(\Pi_2)$ of

$$-\mu_2 \Delta_{yy} w(y) + \alpha w(y) = 1 \quad \text{in } \Pi_2$$

$$w(y) = 0 \quad \text{on } \partial\Pi_2.$$

Then $u(x)$ is the unique solution in $H_0^1(\Omega)$ of

$$-\mu_1 \mathrm{div}_x[\sigma^* \nabla_x u(x)] + \alpha(1 - \beta)u(x) = (1 - \beta)f(x) \quad \text{in } \Omega \tag{6.70}$$

$$u = 0 \quad \text{on } \partial\Omega \tag{6.71}$$

with $\beta = \alpha \int_{\Pi_2} w(y) dy$.

ⓘ Remark 6.5 The solution of the problem (6.40)–(6.41) admits no weak H^1 or strong L^2 limit as in the typical linear homogenization problem (c.f., the case study problem from ▶ Section 4) where the homogenized problem describes a single phase medium with only the macroscale present (no ε!). In contrast, the homogenized problem for the double-porosity problem is given by the system of two boundary value problems so that both phases "survive" in the homogenization limit and they are coupled. Moreover, the high permeability phase described by (6.68) contains only macroscale whereas both the micro- and macroscales are still present in the low permeability phase described by (6.69) since the fast variable y is explicitly present in (6.69).

Examples of Explicit Effective Coefficients: Laminated Structures and 2D Checkerboards

© Springer Nature Switzerland AG 2018
L. Berlyand, V. Rybalko, *Getting Acquainted with Homogenization and Multiscale,*
Compact Textbooks in Mathematics,
https://doi.org/10.1007/978-3-030-01777-4_7

7.1 Effective Conductivity of Laminated Structures

We first consider the case study conductivity problem for laminated structures. In such structures material properties change in one direction only. An example of such structures would be alternating layers of two types of homogeneous materials as depicted in ◘ Figure 7.1. When the number of layers is large and the layers are thin we are in the framework of the homogenization problem (2.5)–(2.6). In the general case, the conductivity of a periodic laminated composite is described by a positive 1-periodic function $\sigma = \sigma(y_1)$, where y_1 is the normal direction to the layers. For simplicity of presentation we consider the isotropic (scalar) conductivity case. Then the cell problem (2.14)–(2.15) becomes

$$-(\sigma(y_1)(\chi_1'(y_1) + 1))' = 0, \quad \chi_1 \text{ is 1-periodic,} \tag{7.1}$$

while $\chi_i \equiv 0$ for $i = 2, \ldots, n$. The effective conductivity tensor can be computed explicitly. Indeed,

$$\chi_1'(y_1) + 1 = \frac{\langle \sigma^{-1} \rangle^{-1}}{\sigma(y_1)}, \tag{7.2}$$

and thus

$$\hat{\sigma}_{11} = \langle \sigma^{-1} \rangle^{-1}, \quad \hat{\sigma}_{ii} = \langle \sigma \rangle, \ i \neq 1, \quad \hat{\sigma}_{ij} = 0, \ i \neq j, \tag{7.3}$$

where $\langle \, \cdot \, \rangle$ denotes the average over the periodicity interval $y_1 \in [0, 1]$. For further reading on laminated structures, we refer to Chapter 9 in [75].

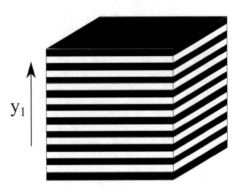

◼ **Fig. 7.1** Laminated material with alternating layers of homogeneous materials

y_1

7.2   Convex Duality and Effective Conductivity of a 2D Checkerboard

In this section we present a beautiful result due to J.B. Keller [64], which was also later obtained independently by A. Dychne [43]. This result was further generalized to nonlinear conductivity problems by O. Levy and R.V. Kohn [67]. In this example we consider a periodic checkerboard geometry of a two-phase composite. While it looks somewhat artificial its further development in Example 8.2 is a quite realistic model of a random mixture of black and white phases with volume fractions p and $1 - p$, respectively.

We first recall elements of the convex duality and the Legendre transform.

Definition 7.1

Let $f : \mathbb{R}^n \to \mathbb{R}^n$ be a convex function with superlinear growth, i.e. $\lim_{|\xi| \to \infty} \frac{|f(\xi)|}{|\xi|} = +\infty$. Define the Legendre transform of f by

$$f^*(\eta) := \sup_{\xi \in \mathbb{R}^n} \{\eta \cdot \xi - f(\xi)\}. \tag{7.4}$$

Clearly this transform can be alternatively defined as $f^*(\eta) = - \inf_{\xi \in \mathbb{R}^n} \{f(\xi) - \eta \cdot \xi\}$. In 1D, the function $f^*(\eta)$ can be understood as the maximal difference between $\eta\xi$ and $f(\xi)$ (see ◼ Figure 7.2). The superlinear growth condition is needed to guarantee that the maximal difference is attained for a finite ξ (i.e., the supremum in (7.4) is attained).

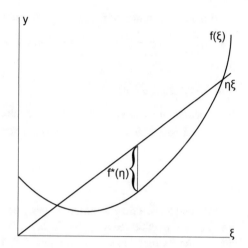

□ Fig. 7.2 Geometric interpretation of the Legendre transform

In order to calculate $f^*(\eta)$ first consider $F(\eta, \xi) := \eta \cdot \xi - f(\xi)$. If there is a vector $\bar{\xi} \in \mathbb{R}^n$ where the supremum of $F(\eta, \cdot)$ is attained, then

$$\nabla_\xi f(\bar{\xi}) = \eta, \tag{7.5}$$

and thus $f^*(\eta) = \eta \cdot \bar{\xi} - f(\bar{\xi})$, or equivalently

$$f(\bar{\xi}) + f^*(\nabla_\xi f(\bar{\xi})) = \bar{\xi}\nabla_\xi f(\bar{\xi}). \tag{7.6}$$

Example 7.1

The following are examples of some useful Legendre transforms.
1. If $f(\xi) = \xi^2$, then $f^*(\eta) = \frac{1}{4}\eta^2$.
2. For $p, q > 1$ with $\frac{1}{p} + \frac{1}{q} = 1$, if $f(\xi) = \frac{|\xi|^p}{p}$, then $f^*(\eta) = \frac{|\eta|^q}{q}$.
3. If $f(\xi) = \frac{1}{2}\sum_{i,j} a_{ij}\xi_i\xi_j$, then $f^*(\eta) = \frac{1}{2}\sum_{i,j} a_{ij}^{(-1)}\eta_i\eta_j$ where the $\{a_{ij}^{(-1)}\}$ are the entries of the inverse matrix of $\{a_{ij}\}$.

∎

Exercise 7.1

Let $f(\xi) = \frac{1}{2}(\xi + \mu) \cdot A(\xi + \mu)$ for $\xi, \mu \in \mathbb{R}^n$ and $A \in \mathbb{R}^{n \times n}$ an invertible matrix. Show that

$$f^*(\eta) = \frac{1}{2}\eta \cdot A^{-1}\eta - \mu \cdot \eta. \tag{7.7}$$

∎

From the definition of the Legendre transform we have

$$F(\eta, \xi) = \eta \cdot \xi - f(\xi) \le f^*(\eta) \quad \forall \eta, \xi \in \mathbb{R}^n, \tag{7.8}$$

thus

$$\eta \cdot \xi \le f(\xi) + f^*(\eta) \quad \forall \eta, \xi \in \mathbb{R}^n. \tag{7.9}$$

In particular, in the case when $f(\xi) = \frac{|\xi|^p}{p}$ and $f^*(\eta) = \frac{|\eta|^q}{q}$ with $p, q > 1$ satisfying $\frac{1}{p} + \frac{1}{q} = 1$ then (7.9) recovers the classical Young's inequality:

$$\xi \cdot \eta \le \frac{|\xi|^p}{p} + \frac{|\eta|^q}{q}. \tag{7.10}$$

We now look at the involution property of the Legendre transform, that is applying it twice to a function recovers this function. It is easy to check the Legendre transform maps a convex function to a convex function. Therefore, it can be iterated on a convex function and it turns out that $f^{**}(x)$ will coincide with the original function $f(x)$.

Proposition 7.1 *If $f \in C^2(\mathbb{R}^n)$ is convex with superlinear growth, then $f^{**}(x) = f(x)$.*

The proof of this classical result can be found in, e.g., [8, 47].

Recall the case study conductivity problem

$$- \operatorname{div}(\sigma(\frac{x}{\varepsilon})\nabla u_\varepsilon(x)) = g(x), \ x \in \Omega \tag{7.11}$$

$$u_\varepsilon(x) = 0, \ x \in \partial\Omega, \tag{7.12}$$

where $\Omega \subset \mathbb{R}^n$ is a bounded domain. The cell problem (2.14)–(2.15) that defines the homogenized problem (2.9)–(2.10) is equivalent to the following minimization problem (cf. Exercise 4.1):

$$\frac{1}{2}\xi \cdot \hat{\sigma}(\xi) = \inf_{v \in V^2_{pot}(\Pi)} \langle (\xi + v) \cdot \frac{1}{2}\sigma(x)(\xi + v) \rangle =: J(\xi), \tag{7.13}$$

where $\Pi = [0, 1]^n$ is the unit periodicity cell, $\langle \cdot \rangle$ is the mean value over Π, and $\sigma(x)$ is a symmetric, invertible, elliptic $n \times n$ matrix on Π. Hereafter we use the subspaces $V^2_{pot}(\Pi)$ and $V^2_{sol}(\Pi)$ defined by

$$V^2_{pot}(\Pi) := \{\nabla u : \ u \in H^1_{per}(\Pi)\},$$

$$V^2_{sol}(\Pi) := \{P \in (L^2(\Pi))^N : \ \langle P \rangle = 0, \quad \langle P, \nabla u \rangle = 0 \ \forall u \in H^1_{per}(\Pi)\},$$

where $H^1_{per}(\Pi)$ is defined in Subsection 1.3.

Recall the Weyl-Helmholtz decomposition:

$$(L^2(\Pi))^n = V^2_{pot}(\Pi) \oplus V^2_{sol}(\Pi) \oplus \mathbb{R}^n. \tag{7.14}$$

Exercise 7.2
Prove the Weyl-Helmholtz decomposition. Hint: Calculate the orthogonal complement $V^2_{pot}(\Pi)^{\perp}$. ∎

Note that (7.13) determines $\hat{\sigma}$. Substituting any potential test function into this (direct) variational formulation provides an upper bound for $\hat{\sigma}$. However, as shown below, $\hat{\sigma}$ is also determined by the following so-called *dual formulation* of the cell problem:

$$\frac{1}{2}\eta \cdot \hat{\sigma}^{-1}\eta = \inf_{p \in V^2_{sol}(\Pi)} \langle (\eta + p) \cdot \frac{1}{2}\sigma^{-1}(x)(\eta + p) \rangle =: D(\eta). \tag{7.15}$$

Using solenoidal test functions in the dual formulation provides upper bounds on $\hat{\sigma}^{-1}$, and thus lower bounds on $\hat{\sigma}$. In the particular case of a two-phase 2D conductivity problem (the checkerboard described below), there is a simple relation between $J(\eta)$ and $D(\eta)$:

$$J(\eta) = \gamma D(\eta) \tag{7.16}$$

where γ is the product of the given conductivities of the phases. Therefore resolving (7.16) with respect to $\hat{\sigma}$ leads to an elegant, explicit formula for $\hat{\sigma}$.

The following convex duality result allows to derive the dual formulation (7.15) from the direct formulation (7.13).

Theorem 7.1 (Convex Duality Theorem for Quadratic Functionals)
Fix $\mu \in \mathbb{R}^n$ and define $f(x, v) := (\mu + v) \cdot \frac{\sigma(x)}{2}(\mu + v)$. For each x denote by $f^*(x, p)$ the Legendre transform of f in v. Let V be any closed subset of $L^2(\Pi)$. Then

$$\inf_{v \in V} \int_{\Pi} f(x, v)dx = -\inf_{p \in V^{\perp}} \int_{\Pi} f^*(x, p)dx, \tag{7.17}$$

where $V^{\perp}$ is the orthogonal complement of V in $L^2(\Pi)$.

Proof
By orthogonality we have $\int_{\Pi} vpdx = 0$ for $v \in V$, $p \in V^{\perp}$. Using this and (7.9) we obtain

$$\int_{\Pi} f(x, v)dx + \int_{\Pi} f^*(x, p)dx = \int_{\Pi} \{f(x, v) + f^*(x, p) - vp\}dx \geq 0, \tag{7.18}$$

from which it follows that

$$\inf_{v \in V} \int_{\Pi} f(x, v)dx \geq \sup_{p \in V^{\perp}} \left(-\int f^*(x, p)dx \right) = -\inf_{p \in V^{\perp}} \int_{\Pi} f^*(x, p)dx. \tag{7.19}$$

Now let $\bar{v}$ be the minimizer of $\int_\Pi f(x, v)dx$ in V and let $\bar{p} := \nabla_p f(x, \bar{v})$. Then $\bar{p} \in V^\perp$ due to the following

$$\frac{\partial}{\partial \varepsilon} \int_\Pi f(x, \bar{v} + \varepsilon w)dx = 0 \Rightarrow \int_\Pi \nabla_p f(x, \bar{v})w dx = 0 \quad \forall w \in V. \tag{7.20}$$

Using (7.6) we obtain

$$\int_\Pi f(x, \bar{v})dx + \int_\Pi f^*(x, \bar{p})dx = \int_\Pi \{f(x, \bar{v}) + f^*(x, \bar{p}) - \bar{v}\bar{p}\}dx = 0. \tag{7.21}$$

From this we get

$$\inf_{v \in V} \int_\Pi f(x, v)dx \leq - \inf_{p \in V^\perp} \int_\Pi f^*(x, p)dx, \tag{7.22}$$

which combined with (7.19) concludes the proof.   □

Using Theorem 7.1, we compute the Legendre transform of $J(\xi)$ to obtain the dual formulation of the cell problem. For simplicity in the following we assume that $\sigma(x)$ is a scalar, that is $\sigma(x) = \sigma(x)I$.

Proposition 7.2 *The Legendre transform of $J(\xi)$ is*

$$D(\eta) := \inf_{p \in V_{sol}^2(\Pi)} \langle (\eta + p) \cdot \frac{1}{2}\sigma^{-1}(x)(\eta + p)\rangle. \tag{7.23}$$

Proof
Let $f(x, v) = (\xi + v) \cdot \frac{1}{2}\sigma(x)(\xi + v)$. Using Exercise 7.1 we see

$$f^*(x, p) = p \cdot \frac{1}{2}\sigma^{-1}(x)p - \xi p.$$

Then Theorem 7.1 and the Weyl-Helmholtz decomposition (7.14) imply

$$J(\xi) = \inf_{v \in V_{pot}^2(\Pi)} \langle (\xi + v) \cdot \frac{1}{2}\sigma(x)(\xi + v)\rangle = - \inf_{p \in (V_{pot}^2(\Pi))^\perp} \langle \frac{1}{2}p \cdot \sigma^{-1}(x)p - \xi p\rangle$$

$$= - \inf_{\lambda \in \mathbb{R}^n} \{ \inf_{p \in V_{sol}^2(\Pi)} \langle \frac{1}{2}(\lambda + p) \cdot \sigma^{-1}(x)(\lambda + p) - \xi(\lambda + p)\rangle \}. \tag{7.24}$$

Since $\rho \in V_{sol}^2(\Pi)$, we have $\langle \rho \rangle = 0$ and

$$J(\xi) = - \inf_{\lambda \in \mathbb{R}^n} \{ \inf_{\rho \in V_{sol}^2(\Pi)} \frac{1}{2}\langle (\lambda + \rho) \cdot \sigma^{-1}(x)(\lambda + \rho)\rangle - \xi\lambda \}. \tag{7.25}$$

Applying the definition of the Legendre transform to $D(\eta)$ yields

$$- \inf_{\lambda \in \mathbb{R}^n} \{D(\lambda) - \xi\lambda\} = D^*(\xi). \tag{7.26}$$

Thus, $J(\xi) = D^*(\xi)$, and taking the Legendre transform of both sides and applying Proposition 7.1 yields $J^*(\eta) = D(\eta)$ as desired. $\qquad\square$

Thus, taking the Legendre transform of both sides of (7.13) we establish the dual cell problem:

$$\eta \cdot \frac{1}{2}\hat{\sigma}^{-1}\eta = D(\eta) = \inf_{p \in V_{sol}^2(\Pi)} \langle (\eta + p)\frac{1}{2}\sigma^{-1}(x)(\eta + p)\rangle. \tag{7.27}$$

Now consider a 2D conductivity problem for a 2-phase medium with a periodic checkerboard pattern. Let $\sigma(x)$ be the conductivity of the periodic checkerboard, see 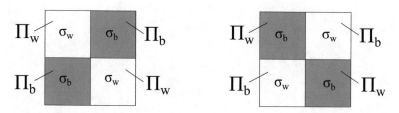 Figure 7.3 (left):

$$\sigma(x) := \begin{cases} \sigma_w & \text{if } x \in \Pi_w \text{ (white phase)} \\ \sigma_b & \text{if } x \in \Pi_b \text{ (black phase)} \end{cases} \tag{7.28}$$

where $\Pi_b = [0, 1/2]^2 \cup [1/2, 1]^2$ is the domain of black squares and $\Pi_w = [0, 1]^2 \setminus \Pi_b$ is the domain of white squares on the checkerboard. Denote by $\widehat{\sigma}(x)$ the conductivity of the recolored checkerboard which is obtained from $\sigma(x)$ by switching values σ_w and σ_b (recoloring), see 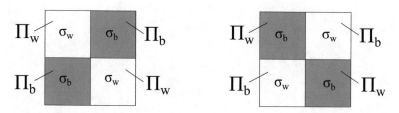 Figure 7.3 (right):

$$\widehat{\sigma}(x) = \begin{cases} \sigma_b & x \in \Pi_w \\ \sigma_w & x \in \Pi_b. \end{cases} \tag{7.29}$$

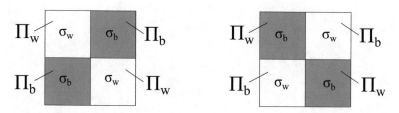

◻ **Fig. 7.3** (left) Original checkerboard pattern, (right) recolored pattern

Exercise 7.3

Show that

$$\widehat{\sigma}(x) = \sigma_b \sigma_w (\sigma(x))^{-1}. \tag{7.30}$$

■

We next present a simple but useful geometric relationship between $V_{pot}^2(\Pi)$ and $V_{sol}^2(\Pi)$ which holds only in 2D. Consider the rotation matrix

$$
R = \begin{bmatrix} 0 & 1 \\ -1 & 0 \end{bmatrix}.
$$

Exercise 7.4
Show that
(i) If $v \in V_{pot}^2(\Pi)$, then $p = Rv \in V_{sol}^2(\Pi)$, and
(ii) if $p \in V_{sol}^2(\Pi)$, then $v = Rp \in V_{pot}^2(\Pi)$.

■

The following exercise shows that the effective conductivity is invariant under recoloring.

Exercise 7.5
Prove that for all $\xi \in \mathbb{R}^2$

$$
\inf_{v \in V_{pot}^2(\Pi)} \langle \widehat{\sigma}(x)|\xi + v|^2 \rangle = \inf_{v \in V_{pot}^2(\Pi)} \langle \sigma(x)|\xi + v|^2 \rangle. \tag{7.31}
$$

Hint: Use a shift by the vector $(1/2, 0)$ to move the cell Π to $\Pi + (1/2, 0)$ and observe how the coloring changes.   ■

Now we are ready to calculate the effective conductivity for the 2D checkerboard conductivity problem.

Theorem 7.2 (Geometric Mean Formula for Effective Conductivity)
The effective conductivity $\widehat{\sigma}$ for the checkerboard pattern $\sigma(x)$ defined in (7.28) is given by the geometric mean:

$$
\widehat{\sigma} = \sqrt{\sigma_b \sigma_w}. \tag{7.32}
$$

Proof
Given $\eta \in \mathbb{R}^2$, let p be the minimizer of (7.27). Set $p = Rv$, $\eta = R\xi$ and use Exercise 7.4 as well as (7.30) to see

$$
\frac{1}{2}\widehat{\sigma}^{-1}\eta^2 = \inf_{p \in V_{sol}^2(\Pi)} \langle (\eta + p) \cdot \frac{1}{2}\sigma^{-1}(\eta + p) \rangle = \inf_{v \in V_{pot}^2(\Pi)} \langle (R\xi + Rv) \cdot \frac{\widehat{\sigma}(x)}{2\sigma_b \sigma_w}((R\xi + Rv)) \rangle
$$

$$= \frac{1}{2\sigma_b\sigma_w} \inf_{v \in V_{pot}^2(\Pi)} \langle \widehat{\sigma}(x)|R(\xi + v)|^2 \rangle.$$

Since $|R(\xi + v)| = |\xi + v|$ we have

$$\frac{1}{2}\hat{\sigma}^{-1}\eta^2 = \frac{1}{2\sigma_b\sigma_w} \inf_{v \in V_{pot}^2(\Pi)} \langle \widehat{\sigma}(x)|\xi + v|^2 \rangle$$

$$\overset{\text{by (7.31)}}{=} \frac{1}{2\sigma_b\sigma_w} \inf_{v \in V_{pot}^2(\Pi)} \langle \sigma(x)|\xi + v|^2 \rangle = \frac{1}{\sigma_b\sigma_w}\left(\frac{1}{2}\hat{\sigma}\xi^2\right) = \frac{1}{\sigma_b\sigma_w}\left(\frac{1}{2}\hat{\sigma}\eta^2\right)$$

$\forall \eta \in \mathbb{R}^2$. Finally we conclude

$$\frac{1}{2}\hat{\sigma}^{-1}\eta^2 = \frac{1}{2\sigma_b\sigma_w}\hat{\sigma}\eta^2 \quad \forall \eta \in \mathbb{R}^2,$$

thus

$$\hat{\sigma} = \sqrt{\sigma_b\sigma_w}.$$

$\square$

The idea of the derivation of (7.32) can be explained as follows. Given two constant conductivities of black and white phases, σ_b and σ_w, we need to compute the (constant) effective conductivity $\hat{\sigma}$ of the homogenized medium. This is done by utilizing the following two relations between these three constants:

- invariance of the effective conductivity under recoloring (Exercise 7.5)
- in 2D switching between direct and dual formulations amounts to rotations of test functions, namely $\hat{\sigma}$ is determined via test functions from $V_{pot}^2(\Pi)$ (see (7.13)) as well as via test functions from $V_{sol}^2(\Pi)$ (see (7.15)).

Introduction to Stochastic Homogenization

© Springer Nature Switzerland AG 2018
L. Berlyand, V. Rybalko, *Getting Acquainted with Homogenization and Multiscale*,
Compact Textbooks in Mathematics,
https://doi.org/10.1007/978-3-030-01777-4_8

While many man-made materials and patterns occurring in nature are well described by periodic models, in various applications one has to introduce mathematical models of stochastic heterogeneous media. Latter models describe heterogeneous materials whose properties cannot be described with certainty and therefore statistical characterization is necessary. Then a specific medium is regarded as an element of a statistical ensemble or, more precisely, a realization of a random field. For instance, we consider below the conductivity problem (8.1)–(8.2) in an inhomogeneous medium whose conductivity tensor is an $n \times n$ random field $A(x/\varepsilon, \omega) = (a_{ij}(y, \omega))_{i,j=1..n}$ where ω is an outcome from Ω, the triple $(\Omega, \mathscr{F}, P)$ being a given probability space, see Subsection 1.8. For a fixed ω from the sample space Ω the function $A(x/\varepsilon, \omega)$ describes the corresponding realization of the random field. Also, solving problem (8.1)–(8.2) for all $\omega \in \Omega$ one obtains yet another random field (process) $u_\varepsilon(x, \omega)$. Both random fields $A(x/\varepsilon, \omega)$ and $u_\varepsilon(x, \omega)$ could be considered as collections of random variables on the probability space $(\Omega, \mathscr{F}, P)$ parametrized by points $x \in G \subset \mathbb{R}^n$. It turns out that under rather general conditions on $A(x/\varepsilon, \omega)$ the random field $u_\varepsilon(x, \omega)$ converges to a deterministic function $u_0(x)$ as $\varepsilon \to 0$, where u_0 is the unique solution of a homogenized problem that does not depend on ε and ω. This conductivity problem is a classical example in the stochastic homogenization theory.

Stochastic homogenization is currently a very active area of research. We restrict our presentation to several basic results and our goal is to introduce the subject of stochastic homogenization rather than present a comprehensive survey. For recent advances in the area we refer the reader to, e.g., [6, 7, 22–24, 27, 30, 31, 53–55] and the references therein.

8.1   Stochastic Case Study Problem and Basic Examples of Models for Random Heterogeneous Media

In this section we consider a stochastic version of the case study conductivity problem and illustrate some basic concepts of stochastic homogenization in the context of this problem. Namely, we study the following linear elliptic PDE in the divergence form:

$$\text{div}\,(A(x/\varepsilon, \omega)\nabla u_\varepsilon(x, \omega)) = g(x), \qquad x \in G \tag{8.1}$$

$$u_\varepsilon(x, w) = 0, \qquad x \in \partial G, \tag{8.2}$$

where $G \subset \mathbb{R}^n$ is a bounded domain (unlike in the previous sections, the PDE is considered in a spatial domain denoted by G as the notation Ω is reserved for the probability space), and $A(y, \omega) = (a_{ij}(y, \omega))_{i,j=1..n}$ is a stationary random field. In other words, at a point x the entries of the tensor $A(x/\varepsilon, \omega)$ are random variables defined on a probability space $(\Omega, \mathscr{F}, P)$ (the definition of probability space and related definitions such as measurable sets, measurable functions, and probability distributions are in Subsection 1.8). The entries $a_{ij}(y, \omega)$ are defined for all points $y \in \mathbb{R}^n$ on the same probability space, and for all $k = 1, 2\ldots$ and $y^{(1)}, y^{(2)}, \ldots, y^{(k)} \in \mathbb{R}^n$ the joint probability distribution of

$$A(y^{(1)} + \xi, \omega),\; A(y^{(2)} + \xi, \omega), \ldots, A(y^{(k)} + \xi, \omega)$$

is independent of $\xi, \forall \xi \in \mathbb{R}^n$. The latter property is the definition of *stationarity* of the field $A(y, \omega)$. Roughly speaking, this property reflects the physical assumption that the microstructure is statistically homogeneous or, in other words, translationally invariant with respect to any shift. This can be compared to the translational invariance of periodic media. In this sense the stationarity property can be viewed as a "stochastic periodicity."
In what follows we consider stationary random fields $A(x, \omega)$ of the form

$$A(y, \omega) = \bar{A}(T(y)\omega), \tag{8.3}$$

where $\bar{A} = \{\bar{a}_{ij}(\omega)\}$ is a given uniformly elliptic and bounded tensor on Ω (measurable with respect to the probability measure P), and $T(y)$ is a *group of measure preserving transformations* (see Definition 8.1). Note that the representation (8.3) holds for *every* stationary random field subject to some natural conditions (such as stochastic continuity and separability) see, e.g., [42], and vice versa, if representation (8.3) holds, then $A(y, \omega)$ is a stationary random field.

Definition 8.1 (Group of Measure Preserving Transformations)

A family of functions $T(y) : \Omega \to \Omega$ parameterized by $y \in \mathbb{R}^n$ is called a *group of measure preserving transformations* if

- $T(0)\omega = \omega$, i.e., $T(0)$ is the identity transformation;
- $T(y + z)\omega = T(y)(T(z)\omega)$;

(continued)

8.1 · Stochastic Case Study Problem and Basic Examples of Models for...

87

8

Definition 8.1 (continued)

- $\forall V \in \mathscr{F}$, $T(y)V \in \mathscr{F}$ and $P[T(y)V] = P[V]$, i.e., $T(y)$ is measure preserving;
- $\forall \bar{f} : \Omega \to \mathbb{R}$ measurable, the function $f(y, \omega) = \bar{f}(T(y)\omega)$ is measurable in $\mathbb{R} \times \Omega$ with respect to the product of the Lebesgue measure in $\mathbb{R}$ and the probability measure P in Ω.

ℹ Remark 8.1 In some homogenization textbooks, e.g., [62], the group of measure preserving transformation $T(y)$ introduced in Definition 8.1 is also referred to as a *dynamical system* with n-dimensional time.

Next we provide three basic examples of models with random fine scale structure. While these examples are very easy to explain on an intuitive level, the main purpose of our presentation is to introduce the reader to rigorous mathematical modeling which allows one to apply powerful probabilistic techniques. Therefore our focus is the identification of the corresponding probability spaces, stationary random fields, and groups of measure preserving transformations.

Example 8.1 (Random Shifts of a Periodic Structure)

In this example we consider shifts of a fixed ε-periodic structure by random vectors. Without loss of generality one can assume that shifts in each coordinate direction are less than the period. Thus such shifts amount to rigid translations of the entire original periodic structure by vectors with small length $O(\varepsilon)$ (microscale). On the other hand homogenization is performed on the macroscale $O(1)$. Therefore it is intuitively clear that the homogenization result for such random shifts is exactly the same as for the original periodic structure.

Next we introduce the probability space Ω corresponding to uniformly distributed random shifts of a periodic structure and define the group of measure preserving transformations $T(y)$. Set $\Omega := \Pi$, where Π is the unit periodicity cell, and consider the Lebesgue measure as a probability measure on Π. Define the group of measure preserving transformations $T(y)$, $y \in \mathbb{R}^n$, by $T(y)\omega = \omega + y - [\omega + y]$, where $\omega \in \Omega$ and $[\,\cdot\,]$ denotes the operation of taking the integer part of each coordinate. Then $T(y)$ captures periodicity of the unit cell, that is if $f(\omega)$ is periodic then $f(T(y)\omega)$ is also periodic. Clearly, for this $T(y)$ conditions of Definition 8.1 hold and $T(y)\omega$ represents random shifts of periodic structure. Consider now a Π-periodic tensor $\sigma(y)$ as in problem (2.5)–(2.6) and define a stationary random field $A(y, \omega)$ by

$$A(y, \omega) := \sigma(T(y)\omega).$$

Since $A(x/\varepsilon, \omega) = \sigma(x/\varepsilon + \omega)$ (thanks to Π-periodicity of σ we have $\sigma(x/\varepsilon + \omega - [x/\varepsilon + \omega]) = \sigma(x/\varepsilon + \omega)$), that is we consider (2.5) with conductivity $\sigma(x/\varepsilon + \omega)$ in place of $\sigma(x/\varepsilon)$ and assume periodicity in $y = x/\varepsilon$. Then solutions of (8.1)–(8.2) converge to the solution of the homogenized problem (2.9)–(2.10) with effective conductivity $\hat{\sigma}(\omega)$ which is actually a constant $\hat{\sigma}$ that does not depend on ω. Indeed, due to periodicity the cell problem (2.14)–

(2.15) is invariant under shifts by ω, therefore neither the existence of the homogenization limit nor coefficients in the homogenized problem are affected by the random shift ω.   ∎

Example 8.2 (Random Checkerboard)

Consider the plane $\mathbb{R}^2$ divided into unit squares with centers in $\mathbb{Z}^2$. Each square is colored black or white with probability p or $1 - p$ independent of each other, and for given positive definite symmetric tensors $\sigma^{(0)}$, $\sigma^{(1)}$ we consider the random field $A(y, \omega)$ that takes values $\sigma^{(0)}$ and $\sigma^{(1)}$ on white and black squares, respectively (◻ Figure 8.1).

◻ **Fig. 8.1** Random checkerboard.

In order to construct the corresponding probability space and the group of measure preserving transformations, introduce first the auxiliary sample space $\Omega' = \{\omega'\} := \{0, 1\}^{\mathbb{Z}^2}$. That is Ω' consists of all possible colorings ω' of the infinite checkerboard. Each coloring is represented by functions from $\mathbb{Z}^2$ to $\{0, 1\}$, where 0 stands for a white square and 1 stands for a black one.

Consider the sets of all colorings with prescribed colors of fixed k squares. Namely, choose arbitrary k different squares with centers $S^{(1)}, S^{(2)}, ..., S^{(k)}$, fix their colors $s^{(1)}, s^{(2)}, ..., s^{(k)}$, and introduce the following subsets of Ω' (also called *cylindrical sets* [94, 95]):

$$B(S, s, k) = \{\omega' \mid \omega'(S^{(1)}) = s^{(1)}, \ldots, \omega'(S^{(k)}) = s^{(k)}\}, \text{ where } S^{(i)} \in \mathbb{Z}^2, \ s^{(i)} \in \{0, 1\}.$$

$$(8.4)$$

Here S and s denote the k-tuples of square centers and their colors. For example, take $k = 1$, $S^{(1)} = (0, 0)$, and $s^1 = 1$, then the set $B(S, s, k)$ consists of all colorings such that the square centered at the origin is black and all other squares are colored arbitrarily. Define the probability P' on these subsets by $P'(B(S, s, k)) = p^{N_b}(1 - p)^{N_w}$, where N_b is the number of black squares ($s^{(i)} = 1$) and N_w is the number of white squares ($s^{(i)} = 0$). Then P' is extended on the σ-algebra $\mathscr{F}'$ generated by all $B(S, s, k)$ that is, the smallest σ-algebra containing all $B(S, s, k)$ (see Subsection 1.8; the extension of P' from the family of sets $B(S, s, k)$ to the σ-algebra is rather technical and can be found in textbooks on probability theory).

Next we introduce the group of measure preserving transformations $T(y)$ generated by *geometrical* shifts of the checkerboard. More precisely, in the case when y has all integer coordinates, $y \in \mathbb{Z}^2$, $T(y)$ is exactly the shift by the vector $-y$. Since the checkerboard is composed of squares whose centers have integer coordinates the shifts with non-integer coordinates are not compatible with the structure of the checkerboard. Thus the definition of $T(y)$ for $y \notin \mathbb{Z}^2$ is more involved. To this end we introduce the auxiliary variable ξ with values in the unit periodicity cell Π, that is $\xi \in \Pi$. Then the corresponding sample space Ω is defined by $\Omega := \Omega' \times \Pi$. One can intuitively think of elements of this space as checkerboards shifted by some $\xi \in \Pi$. Define the probability measure on Ω as the product of the measure P' (on Ω') and the Lebesgue measure (on Π). Now define the group of measure preserving transformations $T(y)$ by setting, for every $\omega = (\omega', \xi) \in \Omega$,

$$T(y)(\omega', \xi) := (\tilde{\omega}, y + \xi - [y + \xi]), \quad \text{where } \tilde{\omega}(z) = \omega'(z + [y + \xi]) \text{ for all } z \in \mathbb{Z}^2.$$

(8.5)

Exercise 8.1
Check that $T(y)$ given by (8.5) satisfies Definition 8.1 of a group of measure preserving transformations. ∎

Finally, setting

$$\bar{A}(\omega) = A = \begin{cases} \sigma^{(0)} \text{ if } \omega'(0) = 0, \\ \sigma^{(1)} \text{ if } \omega'(0) = 1, \end{cases} \quad \text{where } \omega = (\omega', \xi),$$

we define, via (8.3), the stationary field $A(y, \omega)$.

At first glance the random checkerboard sounds somewhat artificial model of two-phase composites because phases in real life are not made out of squares. However if one looks at the random checkerboard from a bird's eye view (far enough not to distinguish individual squares) then one sees black islands randomly distributed in a white ocean or vice versa. Therefore, a random checkerboard can be used as a reasonable model of a heterogeneous medium with random microstructure that has a single microscale.

ⓘ Remark 8.2 If $p = 1/2$ in the above example and $\sigma^{(0)} = \sigma_w I$ and $\sigma^{(1)} = \sigma_b I$ (I is the unit matrix) then we have the same explicit formula for the homogenized conductivity as in the case of the 2D deterministic checkerboard structure described in ▶ Section 7 (see, e.g., Chapter 7 of [62]).

Example 8.3 (Poisson Cloud)

In probability theory, a point process is a collection of points randomly located in a space such as $\mathbb{R}^n$. Specifically, consider the so-called *Poisson point process* or *Poisson cloud*. The elements of the sample space Ω are "point patterns" ω that are fixed distributions of points in $\mathbb{R}^n$ such that every bounded subset on $\mathbb{R}^n$ contains finitely many points. One can think of these elements as of distributions (generalized functions, see Subsection 1.2) of the form

$$\omega(x) = \sum_{j=1}^{\infty} \delta_{z_j}(x), \tag{8.6}$$

where δ_z is the Dirac δ-function centered at z, and z_j are points of the Poisson cloud.

Let us define the probability measure P on Ω. For a given set $S \subset \mathbb{R}^n$ with finite (Lebesgue) measure $|S|$ we introduce a random variable $N_S : \Omega \to \mathbb{Z}_+$ by

$$N_S(\omega) = \int_S \omega(x) dx, \tag{8.7}$$

which is the number of points from a realization ω of the Poisson cloud that are located inside the set S. Consider the event

$$B(S, k) = \{\omega \mid N_S(\omega) = k\}, \tag{8.8}$$

i.e., $B(S, k)$ is the subset of Ω which consists of all realizations ω with exactly k points in S. For each event $B(S, k)$ the probability measure P is defined by

$$P(B(S, k)) = e^{-\lambda|S|} \frac{1}{k!} (\lambda|S|)^k, \tag{8.9}$$

where λ is a positive parameter called the intensity of the process, and k is an arbitrary nonnegative integer. In other words, the (integer valued) random variable N_S (it counts the number of points from ω in S) has the Poisson distribution with the parameter $\lambda|S|$. Moreover, the random variables $N_{S'}$ and $N_{S''}$ are required to be independent for all disjoint sets S' and S''. The probability distribution (8.9) and the requirement of independence (i.e., independence of $N_{S'}$ and $N_{S''}$ for $S' \cap S'' = \emptyset$) uniquely define the point process, called the Poisson cloud. The group of measure preserving transformations $T(y)$ on Ω is given by *geometrical shifts* of the patterns

$$T(y)\omega = \omega(x + y) = \sum_{j=1}^{\infty} \delta_{z_j - y}(x). \tag{8.10}$$

Next for each pattern of points (realization ω of the form (8.6)) we introduce the set

$$F(\omega) = \left\{ y \in \mathbb{R}^n \mid \text{dist}\left(y, \bigcup_i \{z_i\}\right) \leq 1 \right\},$$

91

8

8.1 · Stochastic Case Study Problem and Basic Examples of Models for...

◻ Fig. 8.2 Poisson cloud

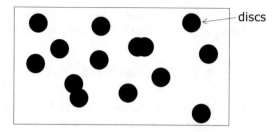

discs

which is the union of unit balls in $\mathbb{R}^n$ centered at these points (the balls may overlap). Then we can introduce the random field $A(y, \omega)$ for all $y \in \mathbb{R}^n$ which describes the two-phase random conductivity (see ◻ Figure 8.2) where balls of conductivity $\sigma^{(0)}$ are randomly distributed in the background media of conductivity $\sigma^{(1)}$:

$$A(y, \omega) = \begin{cases} \sigma^{(0)} & \text{if } y \in F(\omega), \\ \sigma^{(1)} & \text{otherwise.} \end{cases}$$

Here $\sigma^{(0)}$, $\sigma^{(1)}$ are given (deterministic) positive definite symmetric tensors. This random field is stationary. Indeed, it can be equivalently redefined using the group of measure preserving transformations (shifts) (8.10) in (8.3) with

$$\bar{A}(\omega) := \begin{cases} \sigma^{(0)} & \text{if } 0 \in F(\omega), \\ \sigma^{(1)} & \text{otherwise.} \end{cases}$$

∎

Finally, we recall the notion of *ergodic* group of measure preserving transformations. Note that in all three examples above the groups of measure preserving transformations are ergodic.

Definition 8.2 (Ergodic Group of Measure Preserving Transformations)

A group of measure preserving transformations T is called *ergodic* if any measurable function $\bar{f}$ on Ω such that

$$\forall y \in \mathbb{R}^n \quad \bar{f}(T(y)\omega) = \bar{f}(\omega) \text{ a.e. in } \Omega \quad (\bar{f} \text{ is invariant under } T)$$

is constant a.e. in Ω.

Heuristically, this definition can be understood as follows. Applying an ergodic group of measure preserving transformations to any point $\omega \in \Omega$ results in a trajectory $T(y)\omega$ (e.g., if $y \in \mathbb{Z}$, then $T(y)$ corresponds to discrete time iterations $T\omega, T^2\omega, \ldots$) that basically visits the entire space Ω (that is, its trajectory is dense in Ω). Now if a function $\bar{f}$ is invariant under T, then its value will be constant along this trajectory and the density of the trajectory implies that $\bar{f}$ is constant.

8.2   Getting Prepared for Stochastic Homogenization

To state and prove the homogenization theorem for PDEs with random coefficients we need to recall several definitions and classical theorems.

Definition 8.3 (Spatial Average)

Given $f \in L^1_{\text{loc}}(\mathbb{R}^n)$, if the limit

$$\langle f \rangle = \lim_{\rho \to \infty} \frac{1}{\rho^n |K|} \int_{\rho K} f(x) dx \tag{8.11}$$

exists and does not depend on K, it is called a *spatial average* of f. The set K in (8.11) is an arbitrary compact in $\mathbb{R}^n$, and $|K|$ denotes its measure.

Exercise 8.2

Assume that the family of functions $f(x/\varepsilon)$ is bounded in $L^2_{\text{loc}}(\mathbb{R}^n)$ and that the spatial average $\langle f \rangle$ exists. Prove that

$$f(x/\varepsilon) \underset{\varepsilon \to 0}{\rightharpoonup} \langle f \rangle \quad \text{in } L^2_{\text{loc}}(\mathbb{R}^n), \tag{8.12}$$

and vice versa: if (8.12) holds, then the spatial average $\langle f \rangle$ exists. Hint: consider characteristic functions of compact sets K and observe that linear combinations of such functions are dense in $L^2_{\text{loc}}(\mathbb{R}^n)$.   ∎

Note that in the case when a function f is periodic, the spatial average (8.11) exists and equals its mean value over the period due to the Averaging Lemma.

The following important theorem establishes existence and properties of spatial averages for stationary random fields $f(x, \omega) = \bar{f}(T(x)\omega)$. In particular it states that for ergodic systems the spatial average is the same as the average over Ω (the expectation).

Theorem 8.1 (Birkhöff)

Let $\bar{f} \in L^\alpha(\Omega, P)$, $\alpha \geq 1$, and $T(x)$ be a group of measure preserving transformations. Then the stationary random field $\bar{f}(T(x)\omega)$ has a spatial average $\langle \bar{f}(T(x)\omega) \rangle_x$ for almost all $\omega \in \Omega$. Moreover, the mean value is invariant under T as a function of $\omega \in \Omega$, i.e.,

---

(continued)

Theorem 8.1 (continued)

$$\langle \bar{f}(T(x)\omega)\rangle_x = \langle \bar{f}(T(x)T(y)\omega)\rangle_x \qquad \forall y \in \mathbb{R}^n. \tag{8.13}$$

Additionally,

$$E[\bar{f}] = E\left[\langle \bar{f}(T(x)\omega)\rangle_x\right], \tag{8.14}$$

where $E[\cdot]$ denotes the expectation with respect to the probability measure P. Furthermore, if the group of measure preserving transformations T is ergodic, then

$$\langle \bar{f}(T(x)\omega)\rangle_x = E[\bar{f}] \quad \text{for almost all } \omega \in \Omega, \tag{8.15}$$

in particular, $\langle \bar{f}(T(x)\omega)\rangle_x$ does not depend on the realization ω.

🛈 **Remark 8.3** Theorem 8.1 establishes equivalence of spatial and ensemble averages for ergodic systems. In context of time-dependent ergodic systems the analog of the above theorem establishes equivalence of temporal and spatial averages.

Heuristically one can think of the equivalence of temporal and space averages in ergodic systems as follows (S.M. Kozlov, private communication). Consider a piece of material that is subject to heating. Place thermometers at a large number of points in the sample and measure temperature at a particular instant of time and plot a histogram (for temperature distribution). Next place one thermometer at a fixed point and measure temperature for long time to obtain another histogram. Ergodicity in this example means that the average values for both histograms are the same.

Next we present another important result that can be referred to as a "randomized" counterpart of the classical Helmholtz-Weyl decomposition. We begin with definitions of potential and solenoidal (vector-) functions in the framework of stationary vector fields. Note that standard concepts of calculus, e.g., limits, continuity, and derivatives, cannot be defined in a general probability space since there is no notion of distance there. However, if a probability space is supplied with a group of measure preserving transformations, then one can use the theory of dynamical systems to define partial derivatives. In fact, in what follows we only need to define solenoidal and potential vector functions.

--- **Definition 8.4** ---

$\bar{f} \in [L^2(\Omega, P)]^n$ is called *potential* if $f(x, \omega) = \bar{f}(T(x)\omega)$ is potential as a function of x for almost all $\omega \in \Omega$). That is, for almost all ω there exists $U(x, \omega)$, such that $d_x U(x, \omega) = f(x, \omega)$. The space of all potential functions on Ω will be denoted by $L^2_{\text{pot}}(\Omega, P)$.

Definition 8.5

$\bar{f} \in [L^2(\Omega, P)]^n$ is called *solenoidal* if $f(x, \omega) = \bar{f}(T(x)\omega)$ is solenoidal as a function of x for almost all ω. That is, for almost all ω it holds $\operatorname{div}_x f(x, \omega) = 0$. The space of all solenoidal functions on Ω will be denoted by $L^2_{\mathrm{sol}}(\Omega, P)$.

Introducing the spaces

$$V^2_{\mathrm{pot}}(\Omega, P) = \left\{ \bar{f} \in L^2_{\mathrm{pot}}(\Omega, P) \;\middle|\; E[\bar{f}] = 0 \right\}, \tag{8.16}$$

$$V^2_{\mathrm{sol}}(\Omega, P) = \left\{ \bar{f} \in L^2_{\mathrm{sol}}(\Omega, P) \;\middle|\; E[\bar{f}] = 0 \right\}, \tag{8.17}$$

we state

Theorem 8.2   (Helmholtz-Weyl Decomposition)
The following orthogonal decompositions of $[L^2(\Omega, P)]^n$ hold:

$$[L^2(\Omega, P)]^n = V^2_{\mathrm{pot}}(\Omega, P) \oplus L^2_{\mathrm{sol}}(\Omega, P) = L^2_{\mathrm{pot}}(\Omega, P) \oplus V^2_{\mathrm{sol}}(\Omega, P)$$

$$= V^2_{\mathrm{pot}}(\Omega, P) \oplus V^2_{\mathrm{sol}}(\Omega, P) \oplus \mathbb{R}^n. \tag{8.18}$$

The proof is rather technical and can be found in, e.g., Section 7.2 of [62].

8.3   Homogenization Theorem for the Case Study Conductivity Problem in Random Setting

Let $A = (a_{ij}(y, \omega))^n_{i,j=1}$ be a stationary random tensor (matrix) given by (8.3) with a matrix $\bar{A}(\omega) = (\bar{a}_{ij}(\omega))_{i,j=1..n}$ and a group of measure preserving transformations $T(y)$. We assume that $\bar{A}(\omega)$ satisfies for almost all $\omega \in \Omega$ the following boundedness and uniform ellipticity conditions:

$$\bar{a}_{ij}(\omega) \in L^\infty(\Omega) \text{ and } \exists \lambda > 0 \ \lambda|\xi|^2 \leq \xi \cdot \bar{A}(y, \omega)\xi \leq \frac{|\xi|^2}{\lambda} \ \forall \xi \in \mathbb{R}^n, \tag{8.19}$$

where λ does not depend on ω. For simplicity we also assume that the symmetry $\bar{a}_{ij}(y, \omega) = \bar{a}_{ji}(y, \omega)$ holds and the group of measure preserving transformations $T(y)$ is ergodic.

95

8

8.3 · Homogenization Theorem for the Case Study Conductivity Problem in...

Theorem 8.3

There exists a constant (in x and ω) matrix $\hat{A}$ such that, for every given (deterministic) $g \in L^2(G)$ solutions u_ε of (8.1) and their fluxes $A(x/\varepsilon, \omega)\nabla u_\varepsilon$ converge weakly to their homogenized limits $u_0 \in H_0^1(G)$ and $\hat{A}\nabla u_0$ in $H^1(G)$ and $[L^2(G)]^n$, respectively, where u_0 is the unique solution of the homogenized problem

$$\operatorname{div}(\hat{A}\nabla u_0(x)) = g(x), \qquad x \in G \tag{8.20}$$

$$u_0(x) = 0, \qquad x \in \partial G. \tag{8.21}$$

The convergence of solutions and fluxes holds for almost all $\omega \in \Omega$, and u_0 does not depend on ω (it is deterministic). Finally, the homogenized matrix $\hat{A}$ is a symmetric matrix, uniquely defined by

$$\hat{A}\xi \cdot \xi = \min_{\bar{v}_\xi \in V_{pot}^2(\Omega, P)} E[\bar{A}(\xi + \bar{v}_\xi) \cdot (\xi + \bar{v}_\xi)], \quad \forall \xi \in \mathbb{R}^n. \tag{8.22}$$

(See discussion on the computation of the homogenized matrix $\hat{A}$ defined by (8.22) after Exercise 8.3).

ℹ️ **Remark 8.4** The statement of Theorem 8.3 still holds for an arbitrary stationary field $A(y, \omega)$ but without the ergodicity assumption the homogenized problem is no longer deterministic, that is $\hat{A} = \hat{A}(\omega)$.

Proof

The main two steps to prove Theorem 8.3 are

Step I. Introduction of a random counterpart of the cell problem in periodic homogenization. While there is no periodicity in problem (8.1), it turns out that one can introduce an analog of the periodic cell problem which is defined in the entire space and hereafter called the corrector problem. This can be done, thanks to the stationarity property of coefficients that reflects translational invariance (in statistical sense) of random media. A special feature of this corrector problem is that the existence is proved for gradients of the solutions rather than solutions themselves. However, effective conductivity is determined via gradients of solutions just as in periodic problems. The name corrector problem comes from the analogy with the periodic conductivity problem where the corrector u_1 is defined in (4.31) via solution χ_i of the cell problem (2.14)–(2.15).

Step II. Passing to the limit $\varepsilon \to 0$ using Div-Curl Lemma.

Step I (Corrector Problem). In this step we introduce the following stochastic analog of the periodic cell problem:

$$\operatorname{div}_y\Big(A(y, \omega)(\xi + \nabla_y \chi_\xi(y, \omega))\Big) = 0 \qquad y \in \mathbb{R}^n \tag{8.23}$$

with the additional condition

$$\langle \nabla_y \chi_\xi (y, \omega) \rangle_y = 0, \tag{8.24}$$

that replaces the periodic boundary condition. Note that unknown in this problem is $\nabla_y \chi_\xi (y, \omega)$ rather than $\chi_\xi (y, \omega)$ itself and so this problem can be reformulated in terms of $v_\xi (y, \omega) := \nabla_y \chi_\xi (y, \omega)$.

We now show that $v_\xi (y, \omega) = \bar{v}_\xi (T(y)\omega)$ can be defined as the solution of the following problem:

$$\bar{v}_\xi \in V^2_{pot}(\Omega, P), \quad E\left[\phi \cdot \bar{A}(\xi + \bar{v}_\xi)\right] = 0 \quad \forall \phi \in V^2_{pot}(\Omega, P). \tag{8.25}$$

There is a unique solution of this problem. Indeed, rewrite (8.25) as

$$E\left[\phi \cdot \bar{A}\bar{v}_\xi\right] = -E\left[\phi \cdot \bar{A}\xi\right] \tag{8.26}$$

to see that a (unique) solution $\bar{v}_\xi \in V^2_{pot}(\Omega, P)$ exists by the Lax-Milgram Theorem 1.6 using ellipticity of $\bar{A}$. Taking $\phi = \bar{v}_\xi$ in (8.26) and using the Cauchy-Schwarz inequality and ellipticity of $\bar{A}$ and we get

$$E[\bar{v}_\xi \cdot \bar{v}_\xi] < \infty. \tag{8.27}$$

Then by (8.15) in the Birkhoff Theorem, the spatial average is equal to the expectation:

$$\langle \bar{v}_\xi (T(y)\omega) \cdot \bar{v}_\xi (T(y)\omega) \rangle_y = E[\bar{v}_\xi \cdot \bar{v}_\xi] < \infty \text{ for almost all } \omega \in \Omega, \tag{8.28}$$

which in turn yields

$$\bar{v}_\xi (T(x/\varepsilon)\omega) \text{ is bounded in } L^2_{loc}(\mathbb{R}^n). \tag{8.29}$$

This is seen from the following rescaling and the definition of the spatial average

$$\frac{1}{|K|} \int_K |\bar{v}_\xi (T(x/\varepsilon)\omega)|^2 dx = \frac{\varepsilon^n}{|K|} \int_{\frac{1}{\varepsilon}K} |\bar{v}_\xi (T(y)\omega)|^2 dy \xrightarrow[\varepsilon \to 0]{} \langle \bar{v}_\xi (T(y)\omega) \cdot \bar{v}_\xi (T(y)\omega) \rangle_y, \tag{8.30}$$

for any compact set K in $\mathbb{R}^n$. It follows from (8.25) that $\bar{A}(\xi + \bar{v}_\xi)$ belongs to the orthogonal complement to $V^2_{pot}(\Omega, P)$. Then by Theorem 8.2

$$\bar{A}(\xi + \bar{v}_\xi) \in L^2_{sol}(\Omega, P) \quad \forall \xi \in \mathbb{R}^n, \tag{8.31}$$

in particular,

$$\operatorname{div}_y\Big(\bar{A}(T(y)\omega)(\xi + \bar{v}_\xi(T(y)\omega))\Big) = \operatorname{div}_y\Big(A(y,\omega)(\xi + v_\xi(y,\omega))\Big) = 0, \qquad y \in \mathbb{R}^n \tag{8.32}$$

for almost all $\omega \in \Omega$. Therefore, after rescaling $x := \varepsilon y$, we obtain that

$$\operatorname{div}\Big(A(x/\varepsilon,\omega)(\xi + v_\xi(x/\varepsilon,\omega))\Big) = 0 \quad \text{in } \mathbb{R}^n. \tag{8.33}$$

We next use Exercise 8.2 to show that for almost all $\omega \in \Omega$

$$v_\xi(x/\varepsilon,\omega) \rightharpoonup \langle v_\xi(y,\omega)\rangle_y = 0 \quad \text{weakly in } [L^2(\Omega)]^n. \tag{8.34}$$

Indeed, the boundedness in $L^2_{loc}(\mathbb{R}^n)$ follows from (8.29). Ergodicity of $T(y)$ and Theorem 8.1 implies that the spatial average of v_ξ exists and equals $E[\bar{v}_\xi]$. Furthermore since $\bar{v}_\xi \in V^2_{pot}(\Omega, P)$ we have $E[\bar{v}_\xi] = 0$, so

$$\langle \bar{v}_\xi(T(y)\omega)\rangle_y = \langle v_\xi(y,\omega)\rangle_y = 0. \tag{8.35}$$

Note that since $\bar{v}_\xi(T(y)\omega) = v_\xi(y,\omega)$ is a potential (vector-) function, it admits the representation $v_\xi(y,\omega) = \nabla_y \chi(y,\omega)$. Thus (8.23) follows from (8.33) and (8.24) follows from (8.35).

Step II (Homogenization Limit). Consider now solutions u_ε of (8.1) for fixed ω. We assume that ω belongs to a subset $\tilde{\Omega}$ of Ω with $P(\tilde{\Omega}) = 1$ such that curl $v_\xi(T(y)\omega) = 0$ for all $\omega \in \tilde{\Omega}$ and both (8.33) and (8.34) hold for all $\xi \in \mathbb{R}^n$ and all $\omega \in \tilde{\Omega}$. Thanks to the linearity of problem (8.25) in ξ, in order to show existence of such a set $\tilde{\Omega}$ it suffices to consider only n basis vectors $\xi_i = e_i, i = 1, \ldots, n$. Let $\tilde{\Omega}_{\xi_i}$ be a set with $P(\tilde{\Omega}_{\xi_i}) = 1$ such that curl $v_{\xi_i} = 0$ for all $\omega \in \tilde{\Omega}_{\xi_i}$ and both (8.33) and (8.34) hold for all $\omega \in \tilde{\Omega}_{\xi_i}$. Then $\tilde{\Omega} = \bigcap_{i=1}^n \tilde{\Omega}_{\xi_i}$, and $P(\tilde{\Omega}) = 1$.

As in the periodic case (see estimate (2.8)) one establishes an a priori bound which implies (up to a subsequence),

$$u_\varepsilon \rightharpoonup u_0 \text{ in } H^1_0(G), \text{ and } \bar{A}(T(x/\varepsilon)\omega)\nabla u_\varepsilon \rightharpoonup F \text{ in } [L^2(G)]^n. \tag{8.36}$$

Multiplying (8.1) by an arbitrary test function $\varphi \in C^\infty_c(G)$ and integrating over G, we integrate by parts on the left-hand side (cf. weak formulation) and pass to the limit $\varepsilon \to 0$. Using (8.36) to conclude that

$$-\int_G \operatorname{div} F \cdot \nabla\varphi\, dx = \int_G \varphi g\, dx, \tag{8.37}$$

which implies

$$\operatorname{div} F = g. \tag{8.38}$$

The main difficulty now is to relate F to ∇u_0. To this end we use the symmetry of $\bar{A}$ to rewrite the dot product as follows:

$$\bar{A}(T(x/\varepsilon)\omega)\nabla u_\varepsilon \cdot (\xi + v_\xi(T(x/\varepsilon)\omega)) = \bar{A}(T(x/\varepsilon)\omega)(\xi + v_\xi(T(x/\varepsilon)\omega)) \cdot \nabla u_\varepsilon. \tag{8.39}$$

We can pass to the limit as $\varepsilon \to 0$ in (8.39) by applying the Div-Curl Lemma 5.1 to both sides. Indeed for the left-hand side $\operatorname{div}(\bar{A}(T(x/\varepsilon)\omega)\nabla u_\varepsilon) = g$ (is independent of ε) while $\operatorname{curl}(\xi + v_\xi) = 0$ since $v_\xi \in V_{\mathrm{pot}}^2(\Omega)$ and Lemma 5.1 provides the limit $F \cdot \xi$. For the right-hand side of (8.39) we repeat the reasoning in Step I which shows that (8.34) holds to establish the weak L^2 limit

$$\bar{A}(T(x/\varepsilon)\omega)(\xi + \bar{v}_\xi(T(x/\varepsilon)\omega)) \rightharpoonup \langle \bar{A}(T(y)\omega)(\xi + \bar{v}_\xi(T(y)\omega)) \rangle_y. \tag{8.40}$$

By ergodicity of $T(y)$, (8.15) from Theorem 8.1 implies that

$$\bar{A}(T(x/\varepsilon)\omega)(\xi + \bar{v}_\xi(T(x/\varepsilon)\omega)) \rightharpoonup \langle \bar{A}(T(y)\omega)(\xi + \bar{v}_\xi(T(y)\omega)) \rangle_y = E[\bar{A}(\xi + \bar{v}_\xi)]. \tag{8.41}$$

Then using (8.32) and the first convergence in (8.36) with Lemma 5.1 we obtain the limit of the right-hand side of (8.39): $E[\bar{A}(\xi + \bar{v}_\xi)]\rangle \cdot \nabla u_0$. Thus

$$F \cdot \xi = E[\bar{A}(\xi + \bar{v}_\xi)] \cdot \nabla u_0. \tag{8.42}$$

Now, due to the linearity in ξ of $E[\bar{A}(\xi + v_\xi)]$ there exists a constant matrix $\hat{A}$ such that

$$E[\bar{A}(\xi + \bar{v}_\xi)] = \hat{A}\xi \tag{8.43}$$

holds. Combining (8.43) with (8.42) and taking into account the fact that $\xi \in \mathbb{R}^n$ is arbitrary we conclude that $F = \hat{A}\nabla u_0$. Thus (8.38) implies $\operatorname{div}(\hat{A}\nabla u_0) = g$ with $\hat{A}$ defined by (8.43). It is an easy exercise to show that $\hat{A}$ is symmetric, positive definite, and that definition (8.43) is equivalent to (8.22) (Exercise 8.3). Therefore (8.20) has a unique solution u_0 which implies that (8.36) holds for the entire sequence $\{u_\varepsilon\}$ (not just for the subsequence). We thus conclude that u_0 and $\hat{A}\nabla u_0$ are weak limits of u_ε and $\bar{A}(T(x/\varepsilon)\omega)\nabla u_\varepsilon$ as $\varepsilon \to 0$ for almost all $\omega \in \Omega$. $\square$

Exercise 8.3
Prove that if $\hat{A}$ is the homogenized matrix given by (8.43), then $\hat{A}$ is symmetric and positive definite. Moreover show that the definitions (8.43) and (8.22) of $\hat{A}$ are equivalent. Hint: use the symmetry and positivity of $\bar{A}(\omega)$.   ∎

We conclude this subsection with a discussion on the corrector problem (8.23), or equivalently (8.22). The right-hand side of (8.22) looks similar to the energy formulation for the periodic cell problem (4.30). The important distinction is that instead of integration over the periodicity cell in (4.30), the right-hand side of (8.22) contains integration over the entire probability space Ω. That is one moves from the physical space to the probability space. The actual computations are done in the physical space and the ergodicity property of $T(y)$ is used to obtain the expectations which define $\hat{A}$ in (8.22). In order to find the homogenized matrix $\hat{A}$ one has to solve the problem (8.23)–(8.24) for any *typical* realization ω, i.e., for any ω except a subset of Ω with probability 0. For this exceptional ω the homogenization result does not hold. For instance, in the model of random checkerboard from Example 8.2 realizations with all squares having the same color are not typical and cannot be used for calculation of the homogenized tensor. However, in practice any random number generating procedure leads to a typical realization ω.

To solve problem (8.23)–(8.24) one approximates its solutions by those of a problem in a bounded domain. For an accurate approximation this domain should be sufficiently large, e.g., cube $[-L/2, L/2]^n$ with sufficiently large L. Unlike in the periodic case, where homogenized tensor is computed exactly by solving the cell problem in the single periodicity cell, the computational domain in the stochastic case will never be large enough to provide the exact homogenized tensor. This raises the question of the rate of convergence of the approximations as $L \to \infty$. In practical computations this question amounts to the choice of the computational domain which is briefly discussed in the next ▶ Section 8.4.

8.4 Remarks on Recent Developments in Stochastic Homogenization

While classical Theorem 8.3 is a beautiful qualitative result, it does not address the practical issues such as the rate of convergence of u_ε to u_0, also it does not provide a guidance for numerical computation of the homogenized tensor $\hat{A}$. This computation is performed over the so-called *representative volume element* (RVE), that, roughly speaking, is the spatial window which is large enough to represent the microstructure. In the framework of the random conductivity problem this method consists of solving the following truncated corrector problem in a sufficiently large cube $[-L/2, L/2]^n$:

$$\text{div} \, (A(x)(\nabla \phi_{L,\xi} + \xi)) = 0 \quad \text{in } [-L/2, L/2]^n \tag{8.44}$$

for a given realization of conductivity tensor $A(x) = A(x, \omega)$. The problem (8.44) is typically supplemented with periodic boundary conditions. Then, computing the spatial average

$$\hat{A}_L(\omega)\xi = \langle A(x,\omega)(\nabla\phi_{L,\xi}+\xi)\rangle_L := \frac{1}{L^n}\int_{[-L/2,L/2]^n} A(x,\omega)(\nabla\phi_{L,\xi}+\xi)dx,$$

(8.45)

one uses a sampling procedure, e.g., Monte-Carlo method, to generate realizations $\omega = \omega_i$ and find an approximate value for $\hat{A}$ by taking the average

$$\hat{A}_{L,N} = \frac{1}{N}\sum_{i=1}^{N} A_L(\omega_i).$$

(8.46)

From theoretical results $\hat{A}$ is obtained by passing to the limit $N, L \to \infty$, therefore in order to obtain a reliable approximation of $\hat{A}$ by $\hat{A}_{L,N}$ one needs to estimate the rate of convergence.

The question of quantitative estimates of convergence of $A_L(\omega)$ to $\hat{A}$ and $\nabla\phi_{L,\xi}$ to v_ξ as $L \to \infty$ is closely related to estimates on the rate of convergence of solutions u_ε to the homogenized solution u_0 in Theorem 8.3. The latter problem was first addressed in [101], where the rate of convergence of u_ε to u_0 was estimated by $O(\varepsilon^\gamma)$ for some $\gamma > 0$, which was not optimal. In [26] it was shown that for random media in 1D the best exponent is $\gamma = 1/2$, in contrast with the periodic case where the rate of convergence is of the order $\varepsilon^\gamma, \gamma = 1$ (the fact that $\gamma \le 1/2$ in stochastic homogenization follows from the rate of convergence in the central limit theorem). The question of finding the optimal exponent γ in dimensions $n > 1$ is still not completely resolved.

In recent years a substantial progress has been achieved in the development of quantitative methods in stochastic homogenization. In particular, important results for conductivity problem were obtained in the continuum setting and its discrete counterpart when differentiation operations are replaced by finite differences on a lattice. The results in the discrete case are more complete and in order to give an idea of this research direction, we mention here two recent results. It is shown in [55] for i.i.d. coefficients on the lattice that the error estimate for the finite size approximation in $\mathbb{R}^n$ is

$$\left(E[\|\hat{A}_{L,N} - \hat{A}\|^2]\right)^{1/2} \le C\left(\frac{1}{\sqrt{N}}L^{-n/2} + L^{-n}(\log L)^n\right),$$

(8.47)

where $\hat{A}_{L,N}$ is computed via a discrete analogue of (8.46) with N independent realizations $\omega_i, i = 1, ...N$ on the computational domain $[-L/2, L/2]^n$. The rate of convergence in (8.47) is consistent with that of the central limit theorem and is optimal.

The total approximation error in (8.47) is estimated by decomposing it into random and systematic parts. These two parts correspond to two sources of error: the random error due to stochastic fluctuations around the average and the systematic error due to finite size L of the representative volume element (RVE). The random error R_L is defined by the mean square deviation

$$R_L := \sqrt{E[\|\hat{A}_L - E[\hat{A}_L]\|^2]},$$

it measures the deviations of $\hat{A}_L$ from its mean value. The systematic error is given by

$$S_L := \|E[\hat{A}_L] - \hat{A}\|,$$

it provides estimates on the size L of the RVE for the desired accuracy. It is shown in [54] for i.i.d. coefficients and in [55] for more general statistics, that R_L can be estimated by $C(L)L^{-n/2}$ where the pre-factor $C(L)$ is bounded for $n > 2$ and it grows at most logarithmically for $n = 2$. Since $\hat{A}_{L,N}$ is given by (8.46) with independent realizations ω_i, the estimate for R_L leads to the first term in (8.47). It is also shown in [54, 55] that the systematic error S_L decays as $L \to \infty$ almost twice as fast as the random error R_L. These theoretical results allow one to improve the computational efficiency by using empirical averaging (8.46) with $N \approx L^n$, because this choice of N equalizes (to the leading order) the random and the systematic errors.

Γ-Convergence in Nonlinear Homogenization Problems

© Springer Nature Switzerland AG 2018
L. Berlyand, V. Rybalko, *Getting Acquainted with Homogenization and Multiscale*,
Compact Textbooks in Mathematics,
https://doi.org/10.1007/978-3-030-01777-4_9

9.1 What Is Γ-Convergence?

In this section we introduce the Γ-convergence approach to nonlinear problems in a simple 1D setting. There are several textbooks specialized on Γ-convergence which can be recommended for those who want to dive deeper into this important subject, e.g., [28, 40]. Recall the variational form of the case study problem (2.5)–(2.6):

$$\inf_{u \in K} \int_{\Omega} \left[\sigma \left(\frac{x}{\varepsilon} \right) |\nabla u|^2 - 2g(x)u \right] dx, \tag{9.1}$$

where $K = H_0^1(\Omega)$. By equating the first variation of the functional to zero one arrives at the so-called Euler-Lagrange equation for a minimizer. For the quadratic energy functional (9.1) the corresponding PDE problem happens to be linear, and for (9.1) it is nothing but the case study problem (2.5)–(2.6). However, if we consider a non-quadratic energy (which often occurs in physics, materials, and biology), e.g.,

$$\inf_{u \in K} \int_{\Omega} \left[\sigma \left(\frac{x}{\varepsilon} \right) |\nabla u|^p - pg(x)u \right] dx \tag{9.2}$$

where $K = W_0^{1,p}(\Omega)$, $p > 2$, then the resultant Euler-Lagrange equation is nonlinear.

Exercise 9.1
Derive the Euler-Lagrange equation for (9.2) (for example, following the derivation of the Dirichlet principle). ∎

Because of nonlinearity, the previously introduced two-scale expansion and its justification (for example, using the Div-Curl Lemma) may not apply. Therefore we need new mathematical techniques. Such techniques have been developed and they proved to be useful in a wide variety of (nonlinear) problems. They are called the Γ-convergence approach. The main idea is to study convergence of variational functionals and their minimizers rather than solutions of the corresponding PDEs.

These techniques proved to be very powerful since they applied to very general forms of energies (both convex and nonconvex). They also led to beautiful mathematics that is helpful in both analysis and numerics. Moreover, they are more closely related to physics since energy is a basic physical quantity (unlike PDEs).

Consider the following minimization problem:

$$m_\varepsilon = \inf_{u \in K} F_\varepsilon(u), \quad \text{with an integral functional} \quad F_\varepsilon(u) = \int_\Omega \left[f\left(\frac{x}{\varepsilon}, \nabla u\right) - gu \right] dx.$$
$$(9.3)$$

In (9.3) the functional class K describes regularity and boundary conditions of the admissible functions, $g(x)$ is a given "source" function, and $f(y, \lambda)$ is a Lagrangian, which from now on is always assumed Π-periodic in y.

The question we address here is about the asymptotic behavior of minimizers of functionals depending on the small parameter ε, as, for example, integral functionals in (9.3). In this type of variational problems depending on a small parameter a variational convergence introduced by E. De Giorgi called Γ-*convergence* has proved to be an extremely successful concept. The precise definition of Γ-convergence follows.

Definition 9.1

Let $F_\varepsilon(u)$ be a sequence of functionals defined on a metric space (X, d), $F_\varepsilon : X \to \mathbb{R}$. The sequence F_ε Γ-converges to $\hat{F}$, a functional on X, as $\varepsilon \to 0$ if for any $u \in X$:

(i)   (lim inf inequality) For all sequences $\{u^\varepsilon\} \subset X$ satisfying $u^\varepsilon \to u$,

$$\liminf_{\varepsilon \to 0} F_\varepsilon(u^\varepsilon) \geq \hat{F}(u). \tag{9.4}$$

(ii) (lim sup inequality) There exists a sequence $\{u^\varepsilon\} \subset X$ such that for $u^\varepsilon \to u$,

$$\limsup_{\varepsilon \to 0} F_\varepsilon(u^\varepsilon) \leq \hat{F}(u)$$

Such a sequence is called a *recovery sequence*.

Note that a constant sequence $F_\varepsilon(u) = F(u)$ converges to $F(u)$ if and only if the functional $F(u)$ is lower semicontinuous. In this case the lim inf inequality is simply the lower semicontinuity condition (that is, (9.4) is a natural generalization of the lower semicontinuity property), while the lim sup condition is trivially true with $u_\varepsilon = u$. In general, the Γ-limit of the constant sequence $F_\varepsilon = F$ is the lower semicontinuous envelope of F.

Exercise 9.2

Show that the Γ-limit, $\hat{F}$, is always lower semicontinuous. ∎

ⓘ Remark 9.1 Using a recovery sequence $\{u^\varepsilon\}$ for the lim inf inequality (i), we see that

$$\limsup F_\varepsilon(u^\varepsilon) \leq \hat{F}(u) \leq \liminf F_\varepsilon(u^\varepsilon).$$

Not also that if (i) holds then one actually has "=" rather than "≤" in (ii). It is typically harder to prove the lower bound (i) than to show the upper bound (ii). The upper bound requires one to guess a construction of a nearly optimal (lowest energy) sequence of test functions, while establishing the lower bound typically requires to show that this guess of nearly optimal sequence is indeed correct and that the lower bound provided by this guess cannot be improved.

ⓘ Remark 9.2 Note, that both the Γ-limit and Γ-convergence depend on the choice of the metric d. In particular, Γ-limits may differ when strong or weak topology is considered in a variational problem.

There are two main issues which will concern us regarding Γ-convergence and its relation to homogenization theory.
1. How Γ-convergence is related to convergence of minimizers, which is important for many practical problems
2. Establish the compactness property, i.e., for a given sequence of functionals establish existence a Γ-converging subsequence, which implies existence of a homogenization limit for this subsequence.

Definition 9.2

A sequence F_ε is called *equicoercive* on (X, d) if there exists a compact set $K \subset X$ independent of ε such that

$$\inf_{u \in X} F_\varepsilon(u) = \inf_{u \in K} F_\varepsilon(u) \tag{9.5}$$

Note that the equicoercivity of functionals implies the compactness of the sequence of minimizers which allows to establish convergence of minimizers for Γ-convergent functionals. ◻ Figure 9.1 illustrates the notion of equicoercivity for sequence of real valued functions in $\mathbb{R}$.

In the case of integral functionals of the form (9.3), with Π-periodic (in y) Lagrangian $f(y, \lambda)$, a sufficient condition for their equicoercivity in $L^p(\Omega)$, $p > 1$, is the existence of two constants: c, C such that $0 < c < C < \infty$ such that for any $\lambda \in \mathbb{R}^n$,

$$c|\lambda|^p \leq f(y, \lambda) \leq C(1 + |\lambda|^p), \text{ for a.e. } y \in \Pi \text{ (periodicity cell).} \tag{9.6}$$

◘ Fig. 9.1 Equicoercive family of functions

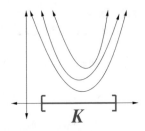

Exercise 9.3

Prove that (9.6) is indeed a sufficient condition of the equicoercivity of the integral functional given by (9.3) in $L^p(\Omega)$. Here Ω is assumed to be a smooth bounded domain in $\mathbb{R}^n$, also g in (9.3) is a function from $L^{p/(p-1)}(\Omega)$. ∎

The relevance of Γ-convergence to homogenization is due to the following property that ensures convergence of minimizers of a Γ-converging sequence of functionals to

Theorem 9.1

Let $F = \Gamma\text{-}\lim F_\varepsilon$ and assume that F_ε is equicoercive in (X, d). Then

1. *(convergence of energies)*

$$\inf_{v \in X} F(v) = \lim_{\varepsilon \to 0} \inf_{v \in X} F_\varepsilon(v); \tag{9.7}$$

2. *(convergence of minimizers) If*

$$u_\varepsilon \in \{\operatorname{argmin}_{v \in X} F_\varepsilon(v)\}, \tag{9.8}$$

and $u_\varepsilon \to u$, then $u \in \{\operatorname{argmin}_{v \in X} F(v)\}$.

Proof

Thanks to the equicoercivity all minimizers $u_\varepsilon = \operatorname{argmin}_{v \in X} F_\varepsilon(v)$ belong to a compact set K. Then, by definition of compact set, the sequence $\{u_\varepsilon\}$ contains a convergent subsequence,

$$u_{\varepsilon'} \to u. \tag{9.9}$$

Using the definition of Γ-convergence and $F_\varepsilon(u_\varepsilon) = \min_{v \in X} F_\varepsilon(v)$, we have

$$F(u) \leq \liminf_{\varepsilon' \to 0} F_{\varepsilon'}(u_{\varepsilon'}) = \liminf_{\varepsilon' \to 0} \inf_{v \in X} F_{\varepsilon'}(v). \tag{9.10}$$

For any $v \in X$, the limsup condition guarantees existence of a recovery sequence $v_{\varepsilon'} \to v$ such that

$$F(v) = \limsup_{\varepsilon' \to 0} F_{\varepsilon'}(v_{\varepsilon'}) \underset{u_{\varepsilon'} \text{ minimizes } F_{\varepsilon'}}{\geq} \limsup_{\varepsilon' \to 0} F_{\varepsilon'}(u_{\varepsilon'}) \geq \liminf_{\varepsilon' \to 0} F_{\varepsilon'}(u_{\varepsilon'}) \underset{(9.10)}{\geq} F(u).$$

$$\tag{9.11}$$

So,

$$F(v) \geq F(u) \tag{9.12}$$

for all $v \in X$ and thus u is a minimizer.

□

Example 9.1 (Γ-Convergence of Quadratic Functionals)

In this example we show that the homogenization theorem 2.1 implies Γ-convergence of the corresponding quadratic functionals. Theorem 9.3 allows for homogenization limits for the case study problem in 1D by Γ-convergence techniques (without appealing to Theorem 2.1).

Consider the functionals

$$F_\varepsilon(u) = \int_\Omega \sigma\left(\frac{x}{\varepsilon}\right) |\nabla u|^2 \, dx. \tag{9.13}$$

We claim that these functionals Γ-converge, with respect to the weak convergence in $H_0^1(\Omega)$, to the functional

$$F(u) = \int_\Omega \hat{\sigma} |\nabla u|^2 \, dx, \tag{9.14}$$

as $\varepsilon \to 0$, where $\hat{\sigma}$ is the homogenized tensor as in Theorem 2.1. Indeed, given u in $H_0^1(\Omega)$, consider solutions $v_\varepsilon \in H_0^1(\Omega)$ of the equation

$$-\mathrm{div}\left(\sigma\left(\frac{x}{\varepsilon}\right)\nabla v_\varepsilon\right) = -\mathrm{div}\left(\hat{\sigma}\nabla u\right) \text{ in } \Omega. \tag{9.15}$$

By Theorem 2.1, $v_\varepsilon \rightharpoonup u$ weakly in $H_0^1(\Omega)$. Moreover, multiplying (9.15) by v_ε and integrating by parts we see that

$$\int_\Omega \nabla v_\varepsilon \cdot \sigma\left(\frac{x}{\varepsilon}\right) \nabla v_\varepsilon \, dx = \int_\Omega \nabla v_\varepsilon \cdot \left(\hat{\sigma}\nabla u\right) dx \to \int_\Omega \nabla u \cdot \left(\hat{\sigma}\nabla u\right) dx. \tag{9.16}$$

This proves the lim sup inequality. Next, consider a sequence $u_\varepsilon \rightharpoonup u$ weakly in $H_0^1(\Omega)$. We have

$$0 \leq \int_\Omega \nabla(u_\varepsilon - v_\varepsilon) \cdot \sigma\left(\frac{x}{\varepsilon}\right) \nabla(u_\varepsilon - v_\varepsilon) \, dx = \int_\Omega \nabla u_\varepsilon \cdot \sigma\left(\frac{x}{\varepsilon}\right) \nabla u_\varepsilon \, dx$$

$$- 2 \int_\Omega \nabla u_\varepsilon \cdot \sigma\left(\frac{x}{\varepsilon}\right) \nabla v_\varepsilon \, dx \tag{9.17}$$

$$+ \int_\Omega \nabla v_\varepsilon \cdot \sigma\left(\frac{x}{\varepsilon}\right) \nabla v_\varepsilon \, dx.$$

By (9.15)

$$\int_{\Omega} \nabla u_{\varepsilon} \cdot \sigma\left(\frac{x}{\varepsilon}\right) \nabla v_{\varepsilon} \, dx = \int_{\Omega} \nabla u_{\varepsilon} \cdot \hat{\sigma} \nabla u \, dx \to F(u), \quad \varepsilon \to 0,$$

therefore taking $\liminf_{\varepsilon \to 0}$ in (9.17) and using (9.16) we arrive exactly in the required inequality (the lim inf condition).   ∎

ⓘ Remark 9.3 The following heuristic explanation of Γ-convergence is useful for developing an intuition. In this explanation the emphasis is on lim inf condition which contains the essence of Γ-convergence. In order to define Γ-limiting functional one starts from an arbitrary function $u(x)$ and considers all the sequences $u_{\varepsilon}(x)$ converging to $u(x)$ as $\varepsilon \to 0$. Then lim inf condition selects an "optimal" sequence $\bar{u}_{\varepsilon}(x)$ with (almost) minimal energies $F_{\varepsilon}(\bar{u}_{\varepsilon})$, that is the inequality in the lim inf condition actually becomes the equality as subsequently verified by the lim sup condition. Finally, the Γ-limiting functional $\hat{F}(u)$ is identified by $\hat{F}(u) = \lim_{\varepsilon \to 0} F_{\varepsilon}(\bar{u}_{\varepsilon})$ for an arbitrary function $u(x)$. Note that for integral functionals of the form (9.3) with fine scale oscillations, the functions in "optimal" sequence $\{u_{\varepsilon}\}_{\varepsilon}$ are constructed locally, i.e., one considers $u(x)$ in some h-neighborhood (where h is a mesoscale, $\varepsilon \ll h \ll 1$) and searches for oscillating approximations $\bar{u}_{\varepsilon} = u(x) + \varepsilon u_1(x, x/\varepsilon)$ by minimizing the integral of the oscillating Lagrangian $f(x/\varepsilon, \nabla \bar{u}_{\varepsilon})$ over this neighborhood. Then u_{ε} is globally constructed by "assembling together the local pieces."

Exercise 9.4
Find Γ-limit of $\{F_{\varepsilon}\}_{\varepsilon}$ given by (9.13) with respect to the strong convergence of $H_0^1(\Omega)$.   ∎

We now state the second main theorem of Γ-convergence theory, a compactness theorem. This theorem states that every sequence of functionals Γ-converges up to a subsequence.

Theorem 9.2 (Γ-Compactness [58])
Let F_{ε} be a sequence of functionals defined on a separable metric space (X, d). Then there exists a Γ-convergent subsequence F_{ε_k} such that

$$\overline{F} = \Gamma\text{-}\lim F_{\varepsilon_k}.$$

9.2 Γ-Convergence for Nonlinear Two-Scale Problems

Consider a two-scale integral functional with a nonlinear Lagrangian $f(y, \lambda)$ that is $\Pi-$periodic in y:

$$F_{\varepsilon}(v) = \int_{\Omega} [f(x/\varepsilon, \nabla v) - g(x)v(x)] \, dx, \quad v \in W_0^{1,p}(\Omega), \ p > 1. \tag{9.18}$$

We make the following assumptions:
1. $g(x) \in L^q(\Omega)$, $p^{-1} + q^{-1} = 1$
2. $\lambda \to f(y, \lambda)$ is continuous, coercive in λ and it satisfies a growth condition:

$$-c_0 + c_1|\lambda|^p \le f(y, \lambda) \le c_0 + c_2|\lambda|^p, \qquad c_0, c_1, c_2 > 0. \tag{9.19}$$

We are looking for a coarse scale functional $\hat{F}(v)$ obtained in the Γ-limit. The natural choice of the functional space here is $W^{1,p}(\Omega)$, however, in order to satisfy the equicoercivity condition (then Γ-convergence implies convergence of minimizers) we consider $W^{1,p}(\Omega)$ endowed with the weak topology since bounded sets in $W^{1,p}(0, 1)$ are weakly compact. It turns out that under an additional convexity assumption on the Lagrangian $f(y, \lambda)$ in λ, the Γ-limit $\hat{F}(v)$ is a coarse scale functional of the form

$$\hat{F}(v) = \int_\Omega [\hat{f}(\nabla v) - g(x)v(x)]dx, \tag{9.20}$$

where the homogenized nonlinear Lagrangian is given by

$$\hat{f}(\lambda) = \inf_{\chi(y) \in W^{1,p}_{per}(\Pi)} \int_\Pi f(y, \lambda + \nabla\chi(y))dy, \qquad \lambda \in \mathbb{R}^n. \tag{9.21}$$

Let us compare (9.21) with the Lagrangian in a special case of quadratic in λ integral functionals corresponding to linear Euler-Lagrange equations. We see that

$$(\hat{\sigma}\lambda, \lambda) = \min_{\chi(y) \in H^1_{per}(\Pi)} \int_\Pi \sigma(y)|\lambda + \nabla_y\chi(y)|^2 dy, \qquad \lambda \in \mathbb{R}^n. \tag{9.22}$$

A comparison of (9.21) and (9.22) illustrates the difference between nonlinear and linear cell problems. Thanks to the linearity of the corresponding Euler-Lagrange equations, (9.22) reduces to n cell problems where $\lambda = e_i$ for $i = 1, ..., n$ (i.e., $\chi_\lambda = \sum \lambda_i \chi_i(y)$). By contrast (9.21) is a one-parameter family of cell problems which, in general, is not reduced to n cell problems.

Note that both (9.21) and (9.22) represent the equivalent energy principle, i.e., the energy of the homogenized medium (the left-hand side of both (9.21) and (9.22)) over the cell Π is equal to the energy of the microstructure in the cell Π.

Suppose now that the Lagrangian $f(y, \lambda)$ is nonconvex in λ. Then the homogenization result becomes more complicated. Namely, an additional minimization over the number of cells appears:

$$\hat{f}(\lambda) = \inf_{k \in N} \inf_{\chi(y) \in W^{1,p}_{per}(k\Pi)} \frac{1}{|k\Pi|} \int_{k\Pi} f(y, \lambda + \nabla\chi(y))dy, \qquad \lambda \in \mathbb{R}^n. \tag{9.23}$$

An example of this phenomenon was observed by S. Müller in [76] while working in elasticity. His example shows that in *vectorial polyconvex* problems minimization over the single periodicity cell ($k = 1$ in (9.23)) is not sufficient for finding the homogenized

Lagrangian $\hat{f}(\lambda)$. Vectorial nature of the problem is essential in that example as for scalar case the homogenized Lagrangian is obtained by solving a single cell problem followed by the convexification (taking a convex envelop, as described in (11.17) in ▶ Section 11.2). Recall that *polyconvexity* is a generalization of convexity for functions whose arguments are matrices (i.e., defined on the space of matrices). A function $f : M^{m \times n} \rightarrow (-\infty, \infty)$ is polyconvex if f is a convex function of all minors of order $1 \leq r \leq \min(m, n)$. More precisely, f is polyconvex if there exists a convex function $g : \mathbb{R}^{\tau(n,m)} \rightarrow \mathbb{R}$ such that

$$f(A) = g(M(A)) \qquad \forall A \in \mathbb{R}^{m \times n}.$$

$M(A)$ is an ordered vector of all minors of order $1, 2, ..., \min(m, n)$, and

$$\tau(m, n) = \sum_{k=1}^{\min(m,n)} \binom{m}{k}\binom{n}{k}.$$

is the total number of all subdeterminates.

Exercise 9.5

1. Show that for $m = n = 1$ (scalar) polyconvexity and convexity are equivalent.
2. For $(m = n = 2)$ find $M(A)$.
3. Sow that, in general, polyconvexity is weaker than convexity by considering $f(A) = \det A$.

■

9.3   Formulation of One-Dimensional Γ-Convergence Theorem

For the sake of simplicity we restrict the study of Γ-convergence of integral functionals with oscillating integrands to the 1D case from [17]. The results presented below can be generalized to multidimensional problems by following essentially the same reasoning.

Definition 9.3

The function $f : \Omega \times \mathbb{R} \rightarrow \mathbb{R}$ $(\Omega \subset \mathbb{R})$ is a function of *Caratheodory type* if
(i)   for every $\lambda \in \mathbb{R}$ the function $y \mapsto f(y, \lambda)$ measurable in Ω.
(ii) For almost all of $y \in \Omega$, $\lambda \mapsto f(y, \lambda)$ is continuous.

Consider the functional

$$F_\epsilon(u) = \begin{cases} \int_0^1 f\left(\dfrac{x}{\varepsilon}, u'(x)\right) dx, & u \in W^{1,p}(0,1) \\ \infty & \text{otherwise,} \end{cases} \tag{9.24}$$

where $p > 1$.

> **ⓘ Remark 9.4** The reader who is not familiar with the notion of a measurable function can assume that $f(y, \lambda)$ is integrable in y. In fact, in this section we consider only integrable functions.

Theorem 9.3
Let the Lagrangian $f(y, \lambda)$ be a Carathéodory type function that satisfies the following:

   (i) $f(y, \xi)$ *is convex in ξ for all $y \in \mathbb{R}$*
   (ii) $f(y, \xi)$ *is 1-periodic in y for all $\xi \in \mathbb{R}$*
   (iii) $\alpha|\xi|^p \leq f(y, \xi) \leq \beta(1 + |\xi|^p)$, *for some $\alpha, \beta > 0$.*

Then the Γ-limit of F_ε as $\varepsilon \to 0$, with respect to the $L^p(0,1)$ convergence, is

$$\Gamma\text{-}\lim_{\epsilon \to 0} F_\epsilon = \hat{F}(u) = \begin{cases} \int_0^1 \hat{f}\left(u'(x)\right) dx, & u \in W^{1,p}(0,1) \\ \infty & \text{otherwise.} \end{cases} \tag{9.25}$$

The homogenized Lagrangian is given by the following nonlinear cell problem:

$$\hat{f}(\lambda) = \min\left\{ \int_0^1 f(y, \chi' + \lambda) dy, \qquad \chi \in W^{1,p}_{per}(0,1) \right\}. \tag{9.26}$$

> **ⓘ Remark 9.5** To explain the ∞ case in (9.25) we note that if function $u \in L^p(0,1)$ is such that $u \notin W^{1,p}(0,1)$, then in order to define the value of functional $F(u)$ one needs to consider smooth approximation of u that is a sequence of functions $u_\varepsilon \in W^{1,p}$ that convergence to u strongly in $L^p(0,1)$. It follows from the coercivity condition (condition (iii) in Theorem 9.3) that values of $F_\varepsilon(u_\varepsilon)$ converge to ∞. Indeed, by contradiction assume that $\{F_\varepsilon(u_\varepsilon)\}$ is bounded. Then $\{u_\varepsilon\}$ is bounded in $W^{1,p}$, i.e., $\|u_\varepsilon\|_{W^{1,p}} < C$ with constant independent from ε. On the other hand, according to Proposition 1.2 weak convergence in $W^{1,p}(0,1)$ implies
> $\liminf_{\varepsilon \to 0} \|u_\varepsilon\|_{W^{1,p}(0,1)} \geq \|u\|_{W^{1,p}(0,1)}$, so since we assumed that $\{u_\varepsilon\}$ is bounded then $\|u\|_{W^{1,p}(0,1)} < \infty$ that contradicts to $u \notin W^{1,p}(0,1)$.

ⓘ **Remark 9.6** The main difference in the nonlinear problem with respect to the linear one is that we get an infinite family of cell problems indexed by a parameter as opposed to the linear case where we always get n cell problems where n is the space dimension.

ⓘ **Remark 9.7** For nonconvex Lagrangians, we replace the condition (i) in the above theorem with

$$(i')\qquad |f(y,\xi) - f(y,\eta)| \leq C(1 + |\xi|^{p-1} + |\eta|^{p-1})|\xi - \eta| \tag{9.27}$$

for every $y \in (0, 1)$ ("uniformly Lipschitz") or its analog in differential condition

$$\left|\frac{\partial f}{\partial \xi}(y,\xi)\right| \leq C(1 + |\xi|^{p-1}). \tag{9.28}$$

There are many physical problems with nonconvex energies, e.g., in phase transitions [12] and theory of crystals [35].

Exercise 9.6
Show that if a Lagrangian $f(y,\xi)$ satisfies conditions (i) and (iii) of Theorem 9.3, then it satisfies (9.27). Show that the inequality (9.27) also holds for the Lagrangian $\hat{f}$ defined by (9.26). ∎

Under the conditions (9.27)–(9.28) in 1D the homogenized Lagrangian $\hat{f}(\lambda)$ is given by the so-called *expanding cell problem*:

$$\hat{f}(\lambda) = \lim_{R\to\infty} \min\left\{\frac{1}{R}\int_0^R f(y, u' + \lambda)dy,\ u(0) = u(R),\ u \in W^{1,p}(0, R)\right\},$$
$$\tag{9.29}$$

while in general multidimensional case the cell problem over expanding domains is given by (9.23).

Exercise 9.7
Suppose dimension $n = 1$. Is (9.23) equivalent to (9.29)? ∎

ⓘ **Remark 9.8** Due to the nonconvexity of the Lagrangian $f(y, \cdot)$ the infimum in χ in problem (9.23) may not be attained for some K. Since the range of K is unbounded the second infimum in K is not necessary attained either. Therefore in order to compute $\hat{f}$ one has to consider sufficiently large K and rigorous definition of $\hat{f}(\lambda)$ requires passing to the limit as $K \to \infty$ (cf. a familiar example from calculus: $\inf_{x\geq 1} 1/x = \lim_{x\to\infty} 1/x$).

Note that Γ-convergence in Theorem 9.3 is considered in the topology of the space $L^p(0, 1)$, while the domain of the functionals F_ε is $W^{1,p}(0, 1)$. The question of Γ-convergence of the functionals F_ε with respect to the (strong) topology of $W^{1,p}(0, 1)$ is much more simple, it leads to the trivial averaged Lagrangian $\langle f(\cdot, \lambda)\rangle$ by applying the Averaging Lemma. However this latter result is useless in the study of minimization problems (9.3), since there is no compactness of minimizers u_ε in the strong topology of $W^{1,p}(0, 1)$.

It should be noted that boundary conditions are not important for the homogenization result as long as the functional F_ε Γ-converges to $\hat{F}$. That is, these results in Theorem 9.3 apply to the Dirichlet problem as well (just replacing $W^{1,p}(0, 1)$ by $W_0^{1,p}(0, 1)$).

We conclude this subsection by attempting to provide intuitive explanation of the cell problem (9.29) (which is equivalent to (9.26) for convex Lagrangians f). First, observe that if homogenized problem exists, then homogenized energy is defined for slowly varying differentiable functions $u \in W^{1,p}(0, 1)$ (more precisely, they are differentiable almost everywhere) which do not depend on ε. Next, we introduce mesoscale h in the homogenized problem such that $\varepsilon \ll h \ll 1$ as in Remark 9.3. The simplest approximation of a function u in the h-neighborhood of a point $x_0 \in (0, 1)$ is its constant value $u(x_0)$. However, energy functionals are defined via derivatives of u and therefore approximations by constants are not sufficient and one has to consider the next order approximation which is linear $u(x) \approx u(x_0) + \lambda(x - x_0)$ ($\lambda = u'(x_0)$). ◻
Figure 9.2 illustrates the continuous piecewise linear approximation of $u(x)$. Rewriting the expanding cell problem (9.29) in the form

$$\lim_{R \to \infty} \min_\chi \frac{1}{R} \int_0^R f(y, (\chi(y) + \lambda y)') \, dy, \tag{9.30}$$

one can see that its solution $\chi(y)$ minimizes the energy of deviations from a linear function with the slope λ, see ◻ Figure 9.3 (right). The minimization problem in (9.30) describes homogenized problem on mesoscale h if one chooses $h = \varepsilon R$. ◻ Figure 9.3 (left) illustrates construction of u_ε in each h-neighborhood as a sum of a linear function with the slope λ and an oscillating corrector:

$$u_\varepsilon(x) \approx u(x_0) + \lambda(x - x_0) + \varepsilon \chi(x/\varepsilon). \tag{9.31}$$

◻ Figure 9.3 (right) illustrates "blown up" profile in coordinates $y = \frac{x - x_0}{\varepsilon}$.

◻ **Fig. 9.2** Construction of piecewise linear approximation with slopes $\lambda_i, i = 1, 2, \ldots$

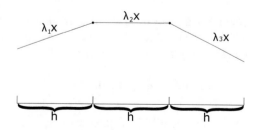

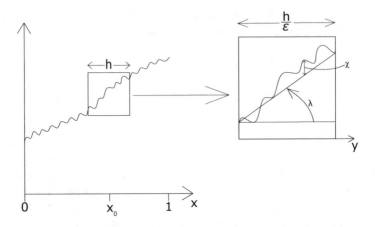

□ **Fig. 9.3** Fine scale description of u_ε; h is the mesoscale ($\varepsilon \ll h \ll 1$); λ is the homogenized (averaged) slope around point x_0. Right figure explains the cell problem $\min_\chi \int_0^1 f(y, (\chi + \lambda y)') \, dy$. Solution of the cell problem χ minimizes the energy of deviation from homogenized solution with slope λ.

9.4   Proof of One-Dimensional Γ-Convergence Theorem

The proof is rather long and we split it into three main parts: A) Reformulation of the Γ-convergence conditions, B) Guessing the appropriate Γ-limit under certain assumptions, C) Establishing Γ-convergence.

Part A: Reformulation of the Γ-Convergence Conditions.   According to the definition of Γ-convergence these two conditions are:

(a)   For every sequence $u_\varepsilon \to u \in L^p(0, 1)$, we have that

$$\liminf_{\varepsilon \to 0} F_\varepsilon(u_\varepsilon) \geq \hat{F}(u). \qquad (\Gamma - \liminf) \tag{9.32}$$

(b)   For all $u \in L^p(0, 1)$, there exists $u_\varepsilon \to u$ in $L^p(0, 1)$ such that

$$\limsup_{\varepsilon \to 0} F_\varepsilon(u_\varepsilon) \leq \hat{F}(u). \qquad (\Gamma - \limsup) \tag{9.33}$$

Instead if checking (a) and (b) it is more convenient to check that the following conditions are satisfied, which turns out to be sufficient for the proof of Theorem 9.3,

(a')   For every sequence $u_\varepsilon \rightharpoonup u$, $u \in W^{1,p}(0, 1)$,

$$\liminf_{\varepsilon \to 0} F_\varepsilon(u_\varepsilon) \geq \hat{F}(u). \tag{9.34}$$

(b')   For all $u \in W^{1,p}(0, 1)$ there exists $u_\varepsilon \rightharpoonup u$ in $W^{1,p}(0, 1)$ such that

$$\limsup_{\varepsilon \to 0} F_\varepsilon(u_\varepsilon) \leq \hat{F}(u). \tag{9.35}$$

Indeed, it is clear that (b')$\Rightarrow$ (b) (since $\hat{F}(u) = \infty$ if $u \notin W^{1,p}(0, 1)$, while for $u \in W^{1,p}(0, 1)$ any weakly converging in $W^{1,p}(0, 1)$ recovery sequence also converges strongly in $L^p(0, 1)$ by the compactness of the embedding $W^{1,p}(0, 1) \subset L^p(0, 1)$). To show that (a')$\Rightarrow$(a) consider an arbitrary sequence $u_\varepsilon \to u$ in $L^p(0, 1)$. Let $\{u_{\varepsilon_j}\}$ be a sequence such that

$$\lim_{j \to \infty} F_{\varepsilon_j}(u_{\varepsilon_j}) = \liminf_{\varepsilon \to 0} F_\varepsilon(u_\varepsilon). \tag{9.36}$$

By coercivity (condition (iii) in Theorem 9.3)

$$F_\varepsilon(u_\varepsilon) = \int_0^1 f\left(\frac{x}{\varepsilon}, u'_\varepsilon(x)\right) dx \geq \alpha \int_0^1 |u'_\varepsilon|^p + \alpha \int_0^1 |u_\varepsilon|^p - \alpha \int_0^1 |u_\varepsilon|^p$$

$$= \alpha \|u_\varepsilon\|_{W^{1,p}}^p - \alpha \|u_\varepsilon\|_{L^p}^p.$$

Consider two cases. First, assume that entire sequence $\{u_\varepsilon\}_\varepsilon$ satisfies $\|u_\varepsilon\|_{W^{1,p}} \to \infty$. It follows that (a) holds in this case since $\liminf_{\varepsilon \to 0} F_\varepsilon(u_\varepsilon) = \infty \geq \hat{F}(u)$. Next, consider the case when there exists a subsequence such that $\|u_{\varepsilon_j}\|_{W^{1,p}} \leq C < \infty$. Then $\{u_{\varepsilon_j}\}_j$ contains a weakly convergent in $W^{1,p}(0, 1)$ subsequence $\{u_{\varepsilon_{j_k}}\}_k$ and its limit must be the same as the strong limit of $\{u_\varepsilon\}_\varepsilon$ in $L^p(0, 1)$ considered in condition (a). Therefore $u_\varepsilon \rightharpoonup u$ weakly in $W^{1,p}(0, 1)$ so that the condition of (a') is satisfied and the lim inf inequality in (a') implies the same inequality in (a).

> **ⓘ Remark 9.9** Note that implications (a)$\Rightarrow$(a') are trivial since weak $W^{1,p}$ convergence implies strong L^p convergence.

Exercise 9.8
Show that (b)$\Rightarrow$(b'). ∎

Part B: Derivation of Nonlinear Cell Problem. The purpose of this part is two-fold: (1) explain the guess for the homogenized Lagrangian $\hat{f}(\lambda)$; (2) show equivalence of periodic cell problem (9.26) and expanding cell problem (9.29). Note that step (1) in this part could be omitted and one can just pass to step (2) followed by proof of part C. Technically, the proof of part C can begin from introducing $\hat{f}(\lambda)$ defined by (9.29) without explaining where it came from.

In this step we proceed with three key assumptions:

(H1) Functionals F_ε Γ-converge to some limit (according to Theorem 9.2, Γ-limit always exists up to a subsequence).

(H2) The Γ-limiting functional has the following integral form:

$$\hat{F}(u) = \int_0^1 \hat{f}(u') dx. \tag{9.37}$$

(H3) $\hat{f}$ is convex.

Assumptions (H1) and (H2) are quite natural if one wishes to derive the form of the homogenized Lagrangian $\hat{f}$. The derivation below also requires the assumption (H3) which is introduced for technical simplicity.

Our goal is to derive from (H1)–(H3) the following nonlinear cell problem that defines $\hat{f}$:

$$\hat{f}(\lambda) = \min \left\{ \int_0^1 f(x, \chi' + \lambda) dx, \quad \chi \in W_{per}^{1,p}(0, 1) \right\}. \tag{9.38}$$

Note that

$$W_{per}^{1,p}(0, 1) = \mathbb{R} + W_0^{1,p}(0, 1) = \{c + v(x) | \; c \in \mathbb{R}, v \in W_0^{1,p}(0, 1)\}.$$

That is one can modify a function $u \in W_{per}^{1,p}(0, 1)$ to satisfy the Dirichlet boundary condition $u(0) = u(1) = 0$ by taking $u - u(0)$ which is in $W_0^{1,p}(0, 1)$ (note that it is possible in 1D but not in higher dimensions). Thus in (9.38) we can consider $\chi \in W_0^{1,p}(0, 1)$. The following identity clearly holds for all $u \in W^{1,p}(0, 1)$ such that $u(0) = u(1) = 0$:

$$\hat{f}(\lambda) = \hat{f}\left(\int_0^1 (\lambda + u') \, dx \right). \tag{9.39}$$

Furthermore due to condition (H3) above, Jensen's inequality applies and it yields

$$\hat{f}(\lambda) \le \int_0^1 \hat{f}(\lambda + u') \, dx \quad \forall u \in W_0^{1,p}(0, 1). \tag{9.40}$$

Exercise 9.9
Show that $\hat{f}$ given by (9.38) satisfies condition (iii) of Theorem 9.3, i.e., $\alpha |\xi|^p \le \hat{f}(\xi) \le \beta(1 + |\xi|^p)$ for $\alpha > 0$ and $\beta > 0$. ∎

Take the minimum in $W_0^{1,p}(0, 1)$ of the right-hand side of (9.40). This minimization converts the inequality from (9.40) into the following equality (e.g., take $u = 0$):

$$\hat{f}(\lambda) = \min \left\{ \int_0^1 \hat{f}(\lambda + u') dx \mid u \in W^{1,p}(0, 1), \; u(0) = u(1) = 0 \right\}. \tag{9.41}$$

In order to derive (9.38) we need to replace $\hat{f}$ by f in the right-hand side of (9.41). To this end we consider a minimizer u_ε of the functional F_ε subject to the boundary conditions:

$$u_{\varepsilon,\lambda}(0) = 0, \qquad u_{\varepsilon,\lambda}(1) = \lambda. \tag{9.42}$$

Exercise 9.10

Show that if conditions (i)–(iii) of Theorem 9.3 are satisfied, then Γ-convergence of the functionals F_ε to $\hat{F}(u) = \int_0^1 \hat{f}(u'(x))dx$ yields the convergence of minimal energies in the Dirichlet problems:

$$F_\varepsilon(u_{\varepsilon,\lambda}) \overset{\varepsilon \to 0}{\to} \min\left\{ \int_0^1 \hat{f}(v')dx \mid v(0) = 0, v(1) = \lambda \right\}$$

$$\overset{u:=v-\lambda x}{=} \min\left\{ \int_0^1 \hat{f}(\lambda + u')\,dx \mid u(0) = u(1) = 0 \right\}. \tag{9.43}$$

This demonstrates that under conditions (i)–(iii) of Theorem 9.3 the boundary conditions are not important for the Γ-limit. Indeed, in Theorem 9.3 we consider Γ-convergence of F_ε in L^p with strong L^p convergence without imposing any boundary conditions (which results in Neumann boundary conditions for minimizers).

Hint: use Exercise 9.6 to prove (9.43). ∎

Thus by (9.43) and (9.41)

$$\hat{f}(\lambda) = \lim_{\varepsilon \to 0} F_\varepsilon(u_{\varepsilon,\lambda}) = \lim_{\varepsilon \to 0} \int_0^1 f\left(\frac{x}{\varepsilon}, u'_{\varepsilon,\lambda}\right)\,dx \tag{9.44}$$

Note that the right-hand side of (9.44) contains oscillations over $\sim \varepsilon^{-1}$ periods. The goal is to obtain the problem (9.38) over single periodicity cell of size 1. To this end consider $\varepsilon_N = \frac{1}{N}$ in (9.44) for integer $N \gg 1$ so that we have N periodicity cells in $(0, 1)$. Setting

$$Nx = y, \qquad u(x) = \frac{1}{N}\tilde{u}(Nx) \tag{9.45}$$

we get

$$\hat{f}(\lambda) = \lim_{N \to \infty} \min\left\{ \int_0^1 f(Nx, u'(x) + \lambda)dx \,\Big|\, u(0) = u(1) = 0 \right\} \tag{9.46}$$

$$= \lim_{N \to \infty} \min\left\{ \frac{1}{N} \int_0^N f(y, \tilde{u}'(y) + \lambda)dy \,\Big|\, \tilde{u}(0) = \tilde{u}(N) = 0 \right\}. \tag{9.47}$$

Next we show that (9.38) and (9.47) define the same $\hat{f}(\lambda)$. To this end observe first that for $N = 1, 2, \ldots$

$$\min\left\{ \frac{1}{N} \int_0^N f(y, \tilde{u}' + \lambda)dy \,|\,\tilde{u}(0) = \tilde{u}(N) = 0 \right\}$$

$$\leq \min\left\{ \int_0^1 f(x, u' + \lambda)dx \,|u(0) = u(1) = 0 \right\}, \tag{9.48}$$

due to the fact that every test function from the right-hand side can be periodically continued to the interval $(0, N)$ and then used in the left-hand side. To prove the converse inequality consider a minimizer of the left-hand side $\tilde{u}$ and construct the 1-periodic test function (to be used in the right-hand side of (9.48))

$$u(x) := \frac{1}{N}\left[\tilde{u}(x) + \tilde{u}(x+1) + \ldots + \tilde{u}(x+(N-1))\right], \tag{9.49}$$

by taking the convex combinations of integer shifts of $\tilde{u}$. Set $v(x) = u(x) - u(0)$. Since $\tilde{u}(0) = \tilde{u}(N)$, we have $v(0) = v(1) = 0$. Furthermore,

$$v'(x) = u'(x) = \frac{1}{N}\left[\tilde{u}'(x) + \tilde{u}'(x+1) + \ldots + \tilde{u}'(x+(N-1))\right]. \tag{9.50}$$

Then, by periodicity of f,

$$\int_0^1 f(x, v'+\lambda)dx = \frac{1}{N}\left[\int_0^1 f(x, u'+\lambda)dx + \cdots + \int_{N-1}^N f(x, u'+\lambda)dx\right]$$

$$= \frac{1}{N}\int_0^N f(x, u'(x)+\lambda)dx. \tag{9.51}$$

Due to convexity of f

$$f(t, u'(x)+\lambda) \le \frac{1}{N}\left[f(x, \tilde{u}'(x)+\lambda) + \ldots + f(x, \tilde{u}'(x+(N-1))+\lambda)\right].$$

Therefore

$$\int_0^1 f(x, v'+\lambda)dx \le \frac{1}{N^2}\sum_{k=0}^{N-1}\int_0^N f(x, \tilde{u}'(x+k)+\lambda)dx = \frac{1}{N}\int_0^N f(x, \tilde{u}'(x)+\lambda)dx. \tag{9.52}$$

Thus formulas (9.38) and (9.47) define the same Lagrangian $\hat{f}(\lambda)$ and we proved that hypotheses (H1)-(H3) imply the formula (9.38) for effective Lagrangian $\hat{f}(\lambda)$.

ⓘ Lemma 9.1 *The following continuous analog of (9.47) holds (switch to a continuous parameter):*

$$\hat{f}(\lambda) = \lim_{R\to\infty} \min\left\{\frac{1}{R}\int_0^R f(x, u'+\lambda)dx \mid u(0) = u(R)\right\}. \tag{9.53}$$

Proof

To estimate the error between (9.47) and (9.53) use the uniform bound $f(x, \xi) \le \beta(1+|\xi|^p)$. Take $N := [R]$ (integer part of R) and choose a minimizing function on the interval $(0, N)$. Then extend it by $u(N)$ on $(0, R)$. Due to the uniform bound on f, we find that

$$\min\left\{\frac{1}{R}\int_0^R f(x, u' + \lambda)dx \mid u(0) = u(R)\right\}$$

$$\leq \min\left\{\frac{1}{N}\int_0^1 f(x, u'+\lambda)dx \mid u(0)=u(N)\right\} + O\left(\frac{1}{N}\right),$$

$$(9.54)$$

then let $N \to \infty$. To obtain the reversed inequality, take a minimizing function on the interval $(0, R)$ and extend it by its value at R to the interval $(R, [R] + 1)$. □

Remark 9.10 Formula (9.41) for a single cell requires computations on the unit interval $(0, 1)$ whereas (9.53) deals with growing intervals $(0, R)$. Therefore, (9.41) is more useful in computations than (9.53). On the other hand, formula (9.53) is more convenient in the proof of Γ-convergence. Moreover, (9.53) holds for the nonconvex case (and its analogues to the random case).

Remark 9.11 Homogenization of a convex energy functional leads to a single cell problem over single periodicity cell whereas nonconvex energies may lead to an "expanding" cell problem over multiples of the basic periodicity cell. This important observation can be explained as follows. First note that for nonconvex energies a minimizer for the single cell problem may not exist. However, even when such a minimizer exists its periodic extension in a nonconvex case is not necessarily the minimizer (only a critical point) on a large mesoscale R (multiple of the basic period).

Exercise 9.11
Establish the following generalization of formula (9.53):

$$\hat{f}(\lambda) = \lim_{R \to \infty} \min\left\{\frac{1}{R}\int_0^R f(x + z, u' + \lambda)dx \,\Big|\, u(0) = u(R)\right\},$$

$$(9.55)$$

for all $z \in \mathbb{R}$. Show also that the limit is uniform in z. ■

Part C: Establishing Γ-Convergence. In this step we show that $\hat{F}$ defined by (9.37)–(9.38) is indeed the Γ-limit of F_ε. Due to part A it is sufficient to prove (a')-(b'). Moreover, it is sufficient to prove (a') and (b') for u from a subset X which is dense in $W^{1,p}(0, 1)$. To show that using a dense subset X is sufficient, we make use of the local Lipschitz continuity of the Lagrangians $f(x, \xi)$ and $\hat{f}(\xi)$ (see Exercise 9.6).

Suppose $u_\varepsilon \rightharpoonup u$ in $W^{1,p}(0, 1)$ and $\tilde{u}_\delta$ is an approximation of u such that $\tilde{u}_\delta \in X$ and

$$\|u - \tilde{u}_\delta\|_{W^{1,p}(0,1)} \leq \delta.$$

$$(9.56)$$

In place of $\{u_\varepsilon\}$ introduce the sequence $\{v_\varepsilon\}$ by

$$v_\varepsilon := u_\varepsilon + \tilde{u}_\delta - u.$$

$$(9.57)$$

Then

$$v_\varepsilon \rightharpoonup \tilde{u}_\delta \text{ as } \varepsilon \to 0, \qquad \text{weakly in } W^{1,p}(0,1). \tag{9.58}$$

We will establish lim inf and lim sup conditions for the sequences $\{v_\varepsilon\}$ weakly converging to elements of X. Then in order to pass from X to $W^{1,p}(0,1)$ it is sufficient to show the following estimate on the energies:

$$|F_\varepsilon(u_\varepsilon) - F_\varepsilon(v_\varepsilon)| \le C\delta, \quad |\hat{F}(u) - \hat{F}(\tilde{u}_\delta)| \le C\delta, \tag{9.59}$$

where C does not depend on δ and ε. To this end, we use (9.27) (established in Exercise 9.6) and Hölder's inequality to get

$$|F_\varepsilon(u_\varepsilon) - F_\varepsilon(v_\varepsilon)| \overset{(9.27)}{\le} C \int_0^1 \left(1 + |u_\varepsilon'|^{p-1} + |v_\varepsilon'|^{p-1}\right) |u_\varepsilon' - v_\varepsilon'| dx$$

$$\overset{\text{Hölder}}{\le} C \|u_\varepsilon' - v_\varepsilon'\|_{L^p} \left(1 + \|u_\varepsilon'\|_{L^p}^{p-1} + \|v_\varepsilon'\|_{L^p}^{p-1}\right)$$

$$\overset{(9.56)}{\le} C \left[1 + \|u_\varepsilon\|_{W^{1,p}}^{p-1} + (\|u_\varepsilon\|_{W^{1,p}} + \delta)^{p-1}\right] \delta \le C\delta,$$

with a constant C independent of u_ε. Similarly, one can prove the following inequality for $\hat{F}$:

$$|\hat{F}(u) - \hat{F}(\tilde{u}_\delta)| \le C \left[1 + \|u\|_{W^{1,p}}^{p-1} + (\|u\|_{W^{1,p}} + \delta)^{p-1}\right] \delta \le C\delta, \tag{9.60}$$

Thus, given $u \in W^{1,p}(0,1)$, a sequence $u_\varepsilon \rightharpoonup u$, and δ-approximation $\tilde{u}_\delta$ of u (in a sense of (9.56)), the estimate (9.59) holds. Therefore it is sufficient to check conditions (a') and (b') on approximating functions $\tilde{u}_\delta$ from a dense subset X in $W^{1,p}(0,1)$. In particular, it is sufficient to prove (a') and (b') on piecewise linear functions. Besides the fact that the set of piecewise linear functions is, in a sense, the simplest dense subset of $W^{1,p}(0,1)$, this choice of the subset X is also quite natural in view of discussion on heuristics in ▶ Section 9.3.

We start from establishing the condition (b') (existence of a recovery sequence) for the piecewise linear functions $u \in X$. Consider a linearity interval (α, β) of $u(x)$, where it is given by

$$u(x) = mx + q, \qquad t \in (\alpha, \beta). \tag{9.61}$$

Motivated by (9.31) we define a recovery sequence $\{u_\varepsilon(x)\}$,

$$u_\varepsilon(x) := mx + q + \varepsilon \chi_\varepsilon \left(\frac{x}{\varepsilon} - \frac{\alpha}{\varepsilon}\right), \qquad t \in (\alpha, \beta), \tag{9.62}$$

where χ_ε is the minimizer of the following problem over the expanding cell:

$$M_\varepsilon := \min_{W^{1,p}} \left\{ \frac{1}{R^\varepsilon} \int_0^{R^\varepsilon} f\left(x + \frac{\alpha}{\varepsilon}, \chi' + m\right) dx \mid \chi(0) = \chi(R_\varepsilon) = 0 \right\} \tag{9.63}$$

with

$$R^\varepsilon = \frac{\beta - \alpha}{\varepsilon}. \tag{9.64}$$

We have shown in Lemma 9.1 (see also Exercise 9.11) that the cell problem (9.38) is equivalent to the expanding cell problem (9.57). Therefore we get (for $\lambda = m$)

$$M_\varepsilon \to \hat{f}(m) \text{ as } \varepsilon \to 0. \tag{9.65}$$

From (9.62), (9.63), and (9.64), we see that

$$\int_\alpha^\beta f\left(\frac{x}{\varepsilon}, u'_\varepsilon(x)\right) dx = \int_\alpha^\beta f\left(\frac{x}{\varepsilon}, m + \chi'_\varepsilon\left(\frac{x}{\varepsilon} - \frac{\alpha}{\varepsilon}\right)\right) dx$$

$$= (\beta - \alpha) M_\varepsilon \xrightarrow{\varepsilon \to 0} (\beta - \alpha)\hat{f}(m).$$

We repeat this construction on each linearity interval of function u. Next, we use the Dirichlet boundary conditions in (9.63) to assemble together function u_ε defined on the entire interval $(0, 1)$. The sequence $\{u_\varepsilon\}$ weakly converges to u due to the above construction and satisfies (b'):

$$\lim_{\varepsilon \to 0} F^\varepsilon(u_\varepsilon) = \hat{F}(u). \tag{9.66}$$

We now verify the condition (a') for piecewise linear functions. As in the proof of (b') above, it suffices to verify the lim sup condition on an interval (α, β) for a given linear function $u = mx + q$. Assume that $u_\varepsilon \rightharpoonup u$ weakly in $W^{1,p}(\alpha, \beta)$. Unlike in the lim sup condition (b'), $\{u_\varepsilon\}$ is now an arbitrary sequence that converges to u and we no longer have an explicit construction like (9.62). In particular, we cannot guarantee that

$$u_\varepsilon(\alpha) = u(\alpha), \qquad u_\varepsilon(\beta) = u(\beta). \tag{9.67}$$

Thus, we need to modify u_ε at the endpoints of the interval (α, β) to approximate $u(\alpha)$ and $u(\beta)$ with a *controlled error* so that the lower bound for energies (9.34) still holds.

ⓘ Remark 9.12 This part of the proof cannot be directly generalized for higher dimensions. A special lemma is proven in [17] to be used in a higher dimensional case.

Thanks to the compact embedding of $W^{1,p}(\alpha, \beta)$ in $C(\alpha, \beta)$, see, e.g., [49], we have the pointwise convergence:

$$u_\varepsilon(\alpha) \to u(\alpha), \qquad u_\varepsilon(\beta) \to u(\beta) \text{ as } \varepsilon \to 0. \tag{9.68}$$

Therefore, we can define a linear function that "corrects" the boundary values:

$$r_\varepsilon(x) := [u(\alpha) - u_\varepsilon(\alpha)]\frac{x - \beta}{\alpha - \beta} + [u(\beta) - u_\varepsilon(\beta)]\frac{x - \alpha}{\beta - \alpha} \tag{9.69}$$

so that $u_\varepsilon + r_\varepsilon = u$ at $x = \alpha, \beta$ and

$$r_\varepsilon \to 0 \text{ strongly in } W^{1,p}(\alpha, \beta). \tag{9.70}$$

From Lipschitz continuity property (9.59) and (9.70), we get

$$\liminf_{\varepsilon \to 0} \int_\alpha^\beta f\left(\frac{x}{\varepsilon}, u'_\varepsilon(x)\right) dx = \liminf_{\varepsilon \to 0} \int_\alpha^\beta f\left(\frac{x}{\varepsilon}, (u_\varepsilon + r_\varepsilon)'(x)\right) dx. \tag{9.71}$$

Next, we rescale the integral in the right-hand side of (9.71) so that (α, β) becomes $(0, R_\varepsilon)$. Then we obtain the expanding cell problem formula:

$$
\begin{aligned}
\int_\alpha^\beta f\left(\frac{x}{\varepsilon}, (u_\varepsilon + r_\varepsilon)'(x)\right) dx &= \int_\alpha^\beta f\left(\frac{x}{\varepsilon}, (u_\varepsilon + r_\varepsilon - mx - q)' + m\right) dx \\
&= \frac{\beta - \alpha}{R_\varepsilon} \int_0^{R_\varepsilon} f\left(y + \frac{\alpha}{\varepsilon}, \chi'_\varepsilon + m\right) dy \\
&\geq (\beta - \alpha) M_\varepsilon,
\end{aligned}
$$

where $R_\varepsilon := \frac{\beta - \alpha}{\varepsilon}$ and $\chi_\varepsilon(y) = \frac{1}{\varepsilon}\tilde{\chi}_\varepsilon(\varepsilon y + \alpha)$ with $\tilde{\chi}_\varepsilon(y) := u_\varepsilon(x) + r_\varepsilon(x) - mx - q$. Thus,

$$\liminf_{\varepsilon \to 0} \int_\alpha^\beta f\left(\frac{x}{\varepsilon}, u'_\varepsilon(x)\right) \geq (\beta - \alpha) \liminf M_\varepsilon = (\beta - \alpha)\hat{f}(m).$$

Thus, (a') is established which completes the proof of Theorem 9.3.

An Example of a Nonlinear Problem: Homogenization of Plasticity and Limit Loads

© Springer Nature Switzerland AG 2018
L. Berlyand, V. Rybalko, *Getting Acquainted with Homogenization and Multiscale*,
Compact Textbooks in Mathematics,
https://doi.org/10.1007/978-3-030-01777-4_10

In this chapter we show how Γ-convergence type techniques can be applied in a specific nonlinear problem. Namely we consider a problem from the theory of plasticity and study convergence of the so-called limit load coefficients which are of primary interest in plasticity theory. This is done by first introducing original and dual formulas for these coefficients followed by constructing sequences of test functions in the spirit of recovery sequences from the classical definition of Γ-convergence.

Roughly speaking plastic deformations are those which lead to irreversible changes (cf. elastic deformations return to the undeformed state once the external force is removed). For example, plastic deformations in metals are caused by dislocations at the atomic scale (defects in structure) [32].

We introduce the notion of plasticity via comparison with elasticity; first recall that exerting force F at the endpoint of an *elastic* spring results in a displacement L (see ◼ Figure 10.1). For a small displacement, the relationship between F and L is linear (see ◼ Figure 10.4):

$$F = kL. \tag{10.1}$$

The relation (10.1) is Hooke's law for an elastic spring where k is the coefficient of stiffness. In higher dimensions we measure deformation of a material using the strain tensor ε. Roughly ε is the infinitesimal change of length of the material relative to the resting length ($\varepsilon = \frac{\Delta L}{L}$). Likewise we measure forces locally using the stress tensor σ.

An *elastic* material (in the linearized regime) then satisfies a constitutive relation called Hooke's law:

$$\sigma = \hat{k}\varepsilon = k\nabla u \tag{10.2}$$

Fig. 10.1 Elasticity

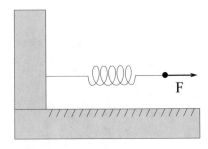

Fig. 10.2 Rigid-perfect plasticity

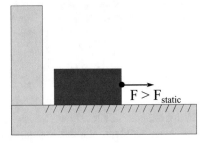

Fig. 10.3 Elasto-perfect plasticity

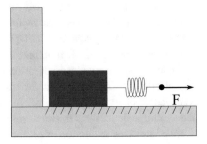

where u is displacement and $\hat{k}$ is called the elastic stiffness tensor (for simplicity we assume that k is a scalar). If g is a fixed external force, then the force balance is given by the equation

$$-\operatorname{div}(k\nabla u) = g. \tag{10.3}$$

The variational formulation of (10.3) is given by minimization of

$$E = \min_{u \in H_0^1(\Omega)} \left\{ \frac{1}{2} \int_\Omega k\,|\nabla u|^2 dx - \int_\Omega gu\,dx \right\}. \tag{10.4}$$

To explain the concept of *plasticity* we first consider a toy example: the dynamics of a mass on a rough surface (**Figure 10.2**, cf. spring). In this case the relation between force and displacement changes: to have a non-zero displacement L, the force F needs to exceed the static friction F_{static}. Once F is slightly larger than F_{static}, one can obtain any displacement by exerting F for a sufficiently long time. Hence, in the corresponding

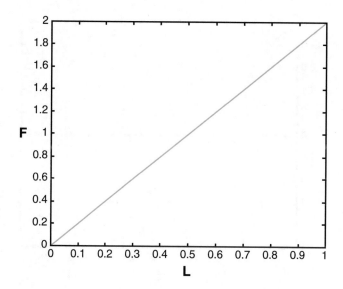

◼ Fig. 10.4 F vs L: Elasticity

constitutive relation the force F is constant in L (see ◼ Figure 10.5). Behavior described by such a relation is called *rigid-perfectly plastic*.

In *rigid-perfect plasticity*, the relation between stress σ and strain ε is

$$\sigma = \hat{k}\frac{\varepsilon}{|\varepsilon|} = k\frac{\nabla u}{|\nabla u|} \tag{10.5}$$

where u is displacement and $\hat{k}$ is the plastic stiffness tensor. Again for simplicity we assume that k is a scalar. Here $\frac{\nabla u}{|\nabla u|}$ has unit length (as expected from the stress-strain relationship) and points in the direction of the stress σ. Again, for external force g, force balance implies that

$$-\mathrm{div}\left(k\frac{\nabla u}{|\nabla u|}\right) = g \tag{10.6}$$

and the corresponding variational formulation is of the form

$$E = \inf_{u \in W^{1,1}(\Omega)}\left\{\int_\Omega f(\nabla u)dx - \int_\Omega gudx\right\}, \quad \text{with Lagrangian } f(\xi) = k|\xi|. \tag{10.7}$$

One can also combine the two above material properties to describe *elasto-perfect plasticity* (see ◼ Figure 10.3). The corresponding constitutive relation is illustrated by ◼ Figure 10.6.

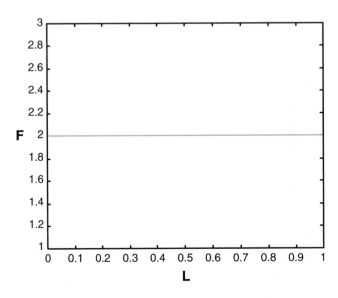

◻ Fig. 10.5 *F* vs *L*: Rigid-perfect plasticity

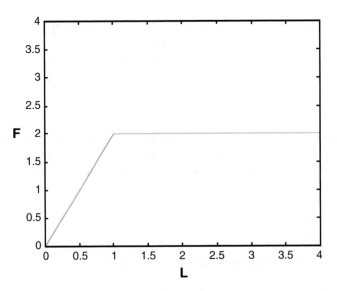

◻ Fig. 10.6 *F* vs *L*: Elasto-perfect plasticity

ⓘ Remark 10.1 In elasto-perfect plasticity, the Lagrangian satisfies the condition that there exist constants $C_0, C_1, C_2 > 0$ such that

$$C_1 |\xi| - C_0 \le f(x, \xi) \le C_2 |\xi| + C_0, \quad x \in \Omega, \ \xi \in \mathbb{R}^n. \tag{10.8}$$

Thus, the main difference between elasticity and plasticity is the form of the Lagrangian. In linear *elasticity* the Lagrangian is quadratic in $|\nabla u|$, while in *rigid-perfect plasticity* it is linear.

10.1 Formulation of Plasticity Problems and Limit Loads

Fix $h \in L^\infty(\Omega)$ and consider the functional $H_h[u]$ given by

$$H_h[u] := \int_\Omega f(x, \nabla u)dx - \int_\Omega hu dx \tag{10.9}$$

where the Lagrangian $f : \Omega \times \mathbb{R} \to \mathbb{R}$ ($\Omega \subset \mathbb{R}$) satisfies the *Caratheodory* condition, i.e., for every $\lambda \in \mathbb{R}$ the function $x \mapsto f(x, \lambda)$ is integrable (in x) and for almost all of $x \in \Omega$, $\lambda \mapsto f(x, \lambda)$ is continuous.

The given function h is the force density applied to the material and unknown u is the resultant displacement of the material: for fixed h we minimize $H_h[u]$ over $u \in W_0^{1,1}(\Omega)$ to obtain the solution.

> ℹ **Remark 10.2** Since the space $W_0^{1,1}(\Omega)$ is not reflexive, minimizers of $H_h[u]$ may or may not exist in this space. One way to avoid this phenomenon is to extend the functional $H_h[u]$ to the space of functions with bounded variation, $BV_0(\Omega)$, which contains at least one minimizer. As our subsequent analysis does not require existence of minimizers we omit these details for simplicity. This consideration can be found in [13].

For simplicity we assume, in addition to (10.8), that the Lagrangian f satisfies the positive homogeneity condition

$$f(x, \lambda\xi) = |\lambda| f(x, \xi), \quad x \in \Omega, \ \xi \in \mathbb{R}^n, \ \lambda \in \mathbb{R}. \tag{10.10}$$

This corresponds to the *rigid-perfectly plastic* case described above. Let $E(h)$ be defined by

$$E(h) := \inf_{v \in W_0^{1,1}(\Omega)} H_h[v]. \tag{10.11}$$

> ℹ **Lemma 10.1** *If f satisfies (10.8) and (10.10), then for any $h \in L^\infty(\Omega)$ we have $E(h)$ is either 0 or $-\infty$.*

Proof

Since $f(x, \xi)$ satisfies (10.10), we have $H_h[\lambda v] = \lambda H_h[v]$ for all $\lambda \in \mathbb{R}^+$ and $v \in W_0^{1,1}(\Omega)$. Given $h \in L^\infty(\Omega)$, consider the two following cases:

Case I: If for all $v \in W_0^{1,1}(\Omega)$ $H_h[v] \geq 0$, we get $0 \leq E(h) \leq H_h[0] = 0$, which implies that $E(h) = 0$.

Case II: If there is $\bar{u} \in W_0^{1,1}(\Omega)$, such that $H_h[\bar{u}] < 0$. Since $H_h[\lambda \bar{u}] = \lambda H_h[\bar{u}]$, by taking the test function $\lambda \bar{u}$ with large λ we obtain $H_h[\lambda \bar{u}] \to -\infty$ as $\lambda \to +\infty$. Thus, $E(h) = -\infty$.

$\square$

Physically, the first case $E(h) = 0$ means that the material perfectly withstands the force density h without any deformations $v = 0$ (e.g., mass on a rough surface does not move). The second case $E(h) = -\infty$ means the material undergoes arbitrarily large deformations $\lambda \bar{u}$, $\lambda \to +\infty$ which are irreversible (e.g., mass on a rough surface may move arbitrarily far). For fixed $p \in L^\infty$, we consider $F(t) := -E(tp)$ as a function of $t > 0$.

Exercise 10.1
Show that $F(t)$ is convex and lower semicontinuous. (**Hint**: use the fact that $-E(tp)$ is defined as a supremum of linear functionals for the load tp). ∎

Lemma 10.1 and Exercise 10.1 imply that there exists $\tau = \tau(p) > 0$, such that $F(t) = 0$ when $t \leq \tau$ and $F(t) = \infty$ when $t > \tau$. In other words for fixed p one may define

$$\tau(p) := \sup\{t > 0 \mid E(tp) > -\infty\}. \tag{10.12}$$

$\tau(p)$ is called *limit load coefficient* and $\tau(p)p$ is called *limit load*.

> **ⓘ Remark 10.3** The limit load depends on the function $p(x)$. One can think of the function p as some fixed force density and consider the family of loads $\{tp(x)\}_{t>0}$. Then the limit load coefficient is the largest t for which the material can perfectly withstand (i.e., $E(tp) \neq -\infty$).

> **ⓘ Remark 10.4** Definition (10.12) is equivalent to

$$\tau(p) = \inf_{\substack{u \in W_0^{1,1}(\Omega) \\ \int_\Omega up=1}} \int_\Omega f(x, \nabla u)dx \tag{10.13}$$

and

$$\frac{1}{\tau(p)} = \sup_{\substack{u \in W_0^{1,1}(\Omega) \\ \int_\Omega f(x,\nabla u)\,dx \leq 1}} \int_\Omega pu\,dx. \tag{10.14}$$

ℹ Lemma 10.2 *Let $\Omega \subset \mathbb{R}^n$ be a bounded domain. There exists $C > 0$ which is independent of p_1, p_2, such that*

$$\left| \frac{1}{\tau(p_1)} - \frac{1}{\tau(p_2)} \right| \leq C \| p_1 - p_2 \|_{L^n(\Omega)} \tag{10.15}$$

for all p_1, $p_2 \in L^\infty(\Omega)$.

Proof

It is clear from (10.14) that $\frac{1}{\tau(p)}$ is convex. According to (10.8), we have

$$0 \leq \frac{1}{\tau(p)} \leq C_1^{-1} \sup_{\substack{u \in W_0^{1,1}(\Omega) \\ \int_\Omega |\nabla u| \, dx \leq 1}} \int_\Omega p u \, dx. \tag{10.16}$$

It follows from the Sobolev embedding $\| u \|_{L^{n/(n-1)}(\Omega)} \leq k \| \nabla u \|_{L^1(\Omega)}$ that

$$0 \leq \frac{1}{\tau(p)} \leq C_3 \| p \|_{L^n(\Omega)}. \tag{10.17}$$

For p_1, $p_2 \in L^\infty(\Omega)$ rewrite $p_2 = \lambda p_1 + (1-\lambda) \left(\frac{p_2}{1-\lambda} - \frac{\lambda p_1}{1-\lambda} \right)$. Using convexity and (10.17):

$$\frac{1}{\tau(p_2)} \leq \lambda \frac{1}{\tau(p_1)} + (1-\lambda) \frac{1}{\tau\left(\frac{p_2}{1-\lambda} - \frac{\lambda p_1}{1-\lambda} \right)} \leq \lambda \frac{1}{\tau(p_1)} + C \| p_2 - \lambda p_1 \|_{L^n(\Omega)}. \tag{10.18}$$

Letting $\lambda \to 1$ we obtain

$$\frac{1}{\tau(p_2)} - \frac{1}{\tau(p_1)} \leq C \| p_2 - p_1 \|_{L^n(\Omega)}. \tag{10.19}$$

Reversing the roles of p_1 and p_2 completes the proof. □

ℹ Remark 10.5 While p is an arbitrary function in $L^\infty(\Omega)$, and we cannot approximate it in the $L^\infty(\Omega)$ metric by smooth functions, Lemma 10.2 shows that τ is continuous with respect to the $L^n(\Omega)$ metric. As such we may use density of smooth functions in $L^n(\Omega)$ for subsequent analysis.

In the sequel we also consider the dual formulation of the limit load problem. Consider the Legendre transform $g(x, \xi)$ of the Lagrangian $f(x, \xi)$ defined by

$$g(x, \xi) = \sup_{\eta \in \mathbb{R}^n} (\xi \cdot \eta - f(x, \eta)), \tag{10.20}$$

where $x \in \mathbb{R}^n$ and $\xi \in \mathbb{R}^n$.

ⓘ Remark 10.6 The definition of the Legendre transform already introduced in ▶ Section 7 requires superlinear growth conditions. In this section we consider the more general context of functions with only linear growth (10.8) so that $g(x, \xi)$ now may take $+\infty$ values.

The function $g(x, \xi)$ is called the conjugate function of $f(x, \xi)$ with respect to ξ. If f satisfies (10.10), then g takes only the values 0 or $+\infty$. For fixed x, a point ξ is said to belong to the domain $\Lambda(x)$ of $g(x, \xi)$ if $g(x, \xi) < +\infty$. Since $g(x, \xi)$ is the Legendre transform of f, then g is a convex, lower semicontinuous function. Therefore its domain $\Lambda(x)$ is a closed convex set, and

$$g(x, \xi) = \begin{cases} 0 & \text{if } \xi \in \Lambda(x), \\ +\infty & \text{if } \xi \notin \Lambda(x). \end{cases} \tag{10.21}$$

The notion of conjugate function is generalized to functionals as follows, and is known as Fenchel-Young-Legendre transform. For a functional $F(u)$ on a Banach space X define $F^*(v)$ for $v \in X^*$ by

$$F^*(v) = \sup_{u \in X}(\langle v, u \rangle - F(u)), \tag{10.22}$$

where $\langle \cdot, \cdot \rangle$ denotes the duality pairing. We use the following theorem to introduce a dual formula for the limit load coefficient.

Theorem 10.1 (Duality Theorem [47])
Let V be a closed convex subset of a Banach space X. Let F be a continuous and convex functional defined on X and F^* be the dual functional of $F(\xi)$ defined on X^*. For any $\xi \in X$ and $q \in X^*$, we have

$$\inf_{x \in V}\{F(x + \xi) - \langle q, x \rangle\} + \inf_{x^* \in V^\perp}\{F^*(x^* + q) - \langle x^*, \xi \rangle\} = \langle q, \xi \rangle. \tag{10.23}$$

Here $V^\perp := \{\eta \in X^* | \langle \eta, x \rangle = 0, \forall x \in V\}$.

ⓘ Remark 10.7 A particular case of this theorem for quadratic functionals F is presented in ▶ Section 7 with $\xi = q = 0$.

Take $\sigma \in L^\infty(\Omega)$ to be a vector field satisfying $-\text{div}\,\sigma = p$, e.g., set $\sigma = \nabla U$ where U solves $-\Delta U = p$ with Dirichlet conditions on the boundary. Apply Theorem 10.1 with $\xi = 0 \in X = L^1(\Omega)$ and $q = t\sigma \in X^* = L^\infty(\Omega)$ to obtain

$$\inf_{u \in W_0^{1,1}(\Omega)} \left\{ \int_\Omega f(x, \nabla u)dx - \int_\Omega tpudx \right\} = \inf_{u \in W_0^{1,1}(\Omega)} \left\{ \int_\Omega f(x, \nabla u)dx - \int_\Omega t\sigma \nabla udx \right\}$$

$$= \inf_{v \in L^1_{pot}(\Omega)} \left\{ \int_\Omega f(x, v)dx - \int_\Omega t\sigma vdx \right\}$$

$$= - \inf_{w \in L^\infty_{sol}(\Omega)} \int_\Omega g(x, w + t\sigma)dx.$$

where $L^1_{pot}(\Omega) = \{u \in L^1(\Omega) \mid \exists v \in W^{1,1}(\Omega) \text{ s.t. } u = \nabla v\}$ and $L^\infty_{sol}(\Omega) = \{u \in L^\infty(\Omega) \mid \operatorname{div} u = 0\}$. Combined with (10.12) expounded as

$$\tau(p) = \sup\{t > 0 \mid \inf_{u \in W_0^{1,1}(\Omega)} \int_\Omega f(x, \nabla u)dx - \int_\Omega tpudx > -\infty\} \tag{10.24}$$

we obtain the *dual formula* of the limit load coefficient

$$\tau(p) := \sup\{t > 0 \mid \inf_{w \in L^\infty_{sol}(\Omega)} \int_\Omega g(x, w + t\sigma)dx < \infty\}. \tag{10.25}$$

10.2 Convergence of Limit Loads in Homogenization Problems

In the previous sections we studied Γ-convergence of energy functionals. In the plasticity problem, the limit load is of primary interest and we focus our attention on the convergence of the limit loads.

Consider the following family of functionals with rapidly oscillating coefficients

$$F^\varepsilon(u) = \int_\Omega f(\frac{x}{\varepsilon}, \nabla u)dx \tag{10.26}$$

and

$$H_h^\varepsilon[v] = F^\varepsilon(u) - \int_\Omega hvdx \tag{10.27}$$

where $f(y, \xi)$ is Π-periodic function in $y \in \Pi := [0, 1]^n$. Analogously to formula (9.21) (see ▶ Section 9) introduce the homogenized Lagrangian by

$$f_{hom}(\xi) = \inf_{v \in W_{per}^{1,1}(\Pi)} \int_\Pi f(y, \xi + \nabla v)dy \tag{10.28}$$

and define

$$H_h^{hom}(v) = \int_\Omega f_{hom}(\nabla v)dx - \int_\Omega hv\,dx. \tag{10.29}$$

Introduce the spaces

$$L_{pot}^1(\Pi) := \{u \in L^1(\Pi): u = \nabla v \text{ for some } v \in W_{per}^{1,1}(\Pi), \ \langle v \rangle = 0\} \tag{10.30}$$

and

$$L_{sol}^\infty(\Pi) := \{z \in L^\infty(\Pi): \langle z \rangle = 0, \ \langle z, \nabla u \rangle = 0 \text{ for all } u \in W_{per}^{1,1}(\Pi)\}, \tag{10.31}$$

and note that by definition

$$(L_{pot}^1(\Pi))^\perp = L_{sol}^\infty(\Pi) \oplus \mathbb{R}^m. \tag{10.32}$$

ⓘ Lemma 10.3 *The conjugate Lagrangian $g_{hom}(\eta)$ of $f_{hom}(\xi)$ is given by*

$$g_{hom}(\eta) := \inf_{z \in L_{sol}^\infty(\Pi)} \int_\Pi g(y, \eta + z(y))dy, \tag{10.33}$$

where $g = f^*$ is defined in (10.20) (or (10.21)).

Proof
By definition (10.20) we compute

$$f_{hom}^*(\eta) = \sup_{\xi \in \mathbb{R}^m} \left\{ \xi \cdot \eta - \inf_{v \in W_{per}^{1,1}(\Pi)} \int_\Pi f(y, \xi + \nabla v)\,dy \right\} \tag{10.34}$$

$$= \sup_{\xi \in \mathbb{R}^m} \left\{ \xi \cdot \eta - \inf_{w \in L_{pot}^1(\Pi)} \int_\Pi f(y, \xi + w)\,dy \right\} \tag{10.35}$$

$$= \sup_{\xi \in \mathbb{R}^m} \left\{ \xi \cdot \eta - \inf_{w \in L_{pot}^1(\Pi)} \int_\Pi f(y, \xi + w) - \eta \cdot w(y)\,dy \right\}, \tag{10.36}$$

where we have used the fact that $\int_\Pi \eta \cdot w(y)dy = 0$ for each $w \in L_{pot}^1$ and constant vector η. Applying Theorem 10.1 we obtain

$$f_{hom}^*(\eta) = \sup_{\xi \in \mathbb{R}^m} \left\{ \inf_{z \in (L_{pot}^1(\Pi))^\perp} \left\{ \int_\Pi g(y, \eta + z(y)) - z(y) \cdot \xi\,dy \right\} \right\}. \tag{10.37}$$

Rewriting $z(y) = \langle z \rangle + \hat{z}(y)$ where $\langle z \rangle \in \mathbb{R}^m$ and $\int_\Pi \hat{z}(y)dy = 0$ we conclude

$$f_{hom}^*(\eta) = \sup_{\xi \in \mathbb{R}^m} \left\{ \inf_{\langle z \rangle \in \mathbb{R}^m} \left\{ \inf_{\hat{z} \in L_{sol}^\infty(\Pi)} \left\{ \int_\Pi g(y, \eta + \hat{z}(y)) - \langle z \rangle \cdot \xi \, dy \right\} \right\} \right\} \tag{10.38}$$

$$= \sup_{\xi \in \mathbb{R}^m} \left\{ \inf_{\langle z \rangle \in \mathbb{R}^m} \{ g_{hom}(\eta + \langle z \rangle) - (\eta + \langle z \rangle) \cdot \xi + \eta \cdot \xi \} \right\} \tag{10.39}$$

$$= \sup_{\xi \in \mathbb{R}^m} \{ \eta \cdot \xi - g_{hom}^*(\xi) \} = g_{hom}(\eta). \tag{10.40}$$

□

Remark 10.8 Note that $g_{hom}(\eta)$ is either $+\infty$ or 0, and $g_{hom}(\eta) = 0$ if and only if there exists a function $z \in L_{sol}^\infty(\Pi)$ such that $\eta + z(y) \in \Lambda(y)$ for almost all y (cf. (10.21)).

We state without proof the following theorem (see, e.g., [17]) which justifies (10.28) as the homogenized Lagrangian of (10.26).

Theorem 10.2
Let $f(y, \xi)$ be a Lagrangian which satisfies (10.8), (10.10). Assume also that f is a $\Pi - periodic$ function in y. Consider f_{hom} and H_h^{hom} defined by (10.28) and (10.29), respectively. Then $H_h^\varepsilon \xrightarrow{\Gamma} H_h^{hom}$ and $F^\varepsilon \xrightarrow{\Gamma} F^{hom}$ in $W_0^{1,1}(\Omega)$ with respect to the strong convergence in $L^1(\Omega)$.

Remark 10.9 This theorem cannot be directly used to study the limiting behavior of limit loads as $\varepsilon \to 0$ due to possible lack of minimizers in the space $W_0^{1,1}(\Omega)$, which is not reflexive.

For fixed $p \in L^\infty(\Omega)$ we define $E^\varepsilon(h)$ and $E^{hom}(h)$ as in (10.11) and $\tau_\varepsilon(p)$ and $\tau_{hom}(p)$ as in (10.12). The following main result establishes the convergence of the limit loads coefficients in the homogenized limit.

Theorem 10.3
Let $f(\frac{x}{\varepsilon}, \xi)$ be a Lagrangian which satisfies (10.8), (10.10). If $p \in L^\infty(\Omega)$, then $\tau_\varepsilon(p) \to \tau_{hom}(p)$ as $\varepsilon \to 0$.

Proof of Theorem 10.3 is based on the following two technical lemmas in which we construct sequence of test functions in the spirit of recovery sequences from the definition of Γ-convergence. We use formula (10.25) with Lemma 10.4 to obtain a lower bound for τ_ε and formula (10.24) with Lemma 10.5 to obtain an upper bound for τ_ε.

ⓘ Lemma 10.4 *Let $v_0(x)$ be a function in $C^\infty(\Omega)$ such that $v_0(x) \in int(dom(g_{hom}))$ for all $x \in \Omega$. There exists a sequence $\{v_\varepsilon\} \subset L^\infty(\Omega)$ with div $v_\varepsilon \rightharpoonup$ div v_0 weakly in $L^\beta(\Omega)$ for all $\beta > 1$ such that*

$$\int_\Omega g\left(\frac{x}{\varepsilon}, v_\varepsilon(x)\right) dx = \int_\Omega g_{hom}(v_0(x))dx = 0. \tag{10.41}$$

ⓘ Remark 10.10 The exact equality (10.41) in this lemma will be used to derive an upper bound on $\tau_\varepsilon(p)$ by utilizing the dual formula (10.25) for limit loads. This allows us to effectively avoid matching lower bounds for the direct problem. Note that there is no limit $\varepsilon \to 0$ on the left-hand side of in (10.41) since $g(x, \xi)$ takes only the values 0 and $+\infty$. In fact, we construct test functions v_ε so that $g(x, v_\varepsilon(x)) = 0$.

Proof

For simplicity we present the proof in two dimensions. The general case follows exactly the same argument replacing 2D objects by their higher dimensional analogues (e.g., triangles appearing in the proof should be replaced by tetrahedrons in the higher dimensional cases).

Consider the domain of g_{hom}, $\Lambda_0 = \{\xi \in \mathbb{R}^2 | g_{hom}(\xi) = 0\}$. Construct a triangulation lying within a compact subset of Λ_0 containing v_0. Denote the vertices of the kth triangle T_k by $\xi^{(i,k)}$, $i = 1, 2, 3$ (**▣** Figure 10.7).

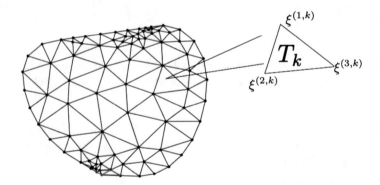

▣ Fig. 10.7 Triangulation of the domain Λ_0

Since each $\xi^{(i,k)}$ belongs to Λ_0 then there exists a minimizer $z^{(i,k)}(y)$ of (10.33) with $\eta = \xi^{(i,k)}$:

$$\int_\Pi g(y, \xi^{(i,k)} + z^{(i,k)}(y))dy = 0. \tag{10.42}$$

We construct a function $Z(y, \xi)$ such that for every $\xi \in \Lambda_0$ we have $Z(\cdot, \xi) \in L^\infty_{sol}(\Pi)$ and $Z(y, \xi) + \xi \in \Lambda(y)$ (see (10.21)). We represent each $\xi \in T_k$ as a convex combination of vertices $\xi^{(i,k)}$, $\xi = \sum_{i=1}^3 \alpha_i \xi^{(i,k)}$ and set

$$Z(y, \xi) := \sum_{i=1}^{3} \alpha_i z^{(i,k)}(y). \tag{10.43}$$

Then due to convexity of $g(y, \xi)$

$$0 \le \int_{\Pi} g(y, \xi + Z(y, \xi)) dy \le \sum_{i=1}^{3} \alpha_i \int_{\Pi} g(y, \xi^{(i,k)} + z^{(i,k)}(y)) dy = 0. \tag{10.44}$$

Since $z^{(i,k)} \in L_{sol}^{\infty}(\Pi)$ (see (10.33)), the construction (10.43) guarantees that $\mathrm{div}_y\, Z(y, \xi) = 0$ and that $Z(y, \xi)$ is Π-periodic in y for each ξ. Also $Z(y, \xi)$ is piecewise linear and continuous in ξ for each y. So $Z(y, \xi)$ is a solution of the minimization problem (10.33). Since $g(y, \xi + Z(y, \xi)) = 0$ for almost every $y \in \Pi$ then

$$\int_{\Omega} g(\frac{x}{\varepsilon}, v_0(x) + Z(\frac{x}{\varepsilon}, v_0(x))) dx = 0 \tag{10.45}$$

and thus (10.41) follows by taking $v_\varepsilon := v_0(x) + Z(\frac{x}{\varepsilon}, v_0(x))$. The construction of Z immediately implies that

$$\mathrm{div}\, v_\varepsilon = \mathrm{div}\, v_0 + \sum_{i,j} \frac{\partial Z_i}{\partial \xi_j} \cdot \frac{\partial (v_0)_j}{\partial x_i} =: \mathrm{div}\, v_0 + \gamma(x, \frac{x}{\varepsilon}), \tag{10.46}$$

and since $Z(y, \xi)$ is piecewise linear in ξ we have

$$\gamma(x, y) = \sum_{i,k} \ell^{(i,k)}(x) \cdot Z(y, \xi^{(i,k)}) \tag{10.47}$$

for piecewise constant vector functions $\ell^{(i,k)}(x) \in L^{\infty}(\Omega)$. Thanks to this representation in separated variables and the fact that $\langle Z(\cdot, \xi) \rangle = 0$ since $Z(\cdot, \xi) \in L_{sol}^{\infty}(\Pi)$, we can use the Averaging Lemma 2.1 to conclude that

$$\gamma(x, \frac{x}{\varepsilon}) \rightharpoonup 0 \tag{10.48}$$

weakly in $L^{\beta}(\Omega)$ for all $\beta > 1$. $\qquad\square$

ⓘ Lemma 10.5 *Let $u_0(x)$ be a function in $C_c^{\infty}(\Omega)$. For all $\delta > 0$ there exists a sequence $\{u_\varepsilon\} \subset W_0^{1,1}(\Omega)$ with $u_\varepsilon \to u_0$ in $L^1(\Omega)$ such that*

$$\limsup_{\varepsilon \to 0} \int_{\Omega} f(\frac{x}{\varepsilon}, \nabla u_\varepsilon) dx \le \int_{\Omega} f_{hom}(\nabla u_0) dx + \delta. \tag{10.49}$$

Proof

As in the proof of Lemma 10.4 we assume for simplicity that the dimension $n = 2$ in this proof. The general case follows exactly the same argument replacing 2D objects by their higher dimensional analogues.

Fix $\delta > 0$ and triangulate $\mathbb{R}^2$ with sufficiently small mesh to ensure the maximal diameter d_T of the triangulation satisfies $d_T \leq \frac{\delta}{2|\Omega|} \cdot \frac{1}{3C_3}$, where C_3 is the Lipschitz constant for f_{hom} (i.e., $|f_{hom}(\xi) - f_{hom}(\eta)| \leq C_3 |\xi - \eta|$). We denote the vertices of the kth triangle T_k by $\xi^{(i,k)}$. According to the definition of f_{hom}, for $\xi^{(i,k)}$ there exists $v^{(i,k)}(y) \in C^\infty_{per}(\Pi)$ such that

$$\int_\Pi f(y, \xi^{(i,k)} + \nabla_y v^{(i,k)}(y))dy \leq f_{hom}(\xi^{(i,k)}) + \frac{\delta}{2|\Omega|}. \tag{10.50}$$

Since $v^{(i,k)}(y)$ is defined up to a constant, we can normalize $v^{(i,k)}$ so that $\int_\Pi v^{(i,k)}(y)dx = 0$. Without loss of generality we assume that $\xi^{(0,0)} = 0$ is a vertex of the triangulation and that $v^{(0,0)}(y) \equiv 0$. As in the proof of Lemma 10.4 we define $N(y, \xi)$ by

$$N(y, \xi) := \sum_{i=1}^{3} \alpha_i v^{(i,k)}(y), \tag{10.51}$$

where $\xi \in T_k$ is represented by a convex combination: $\xi = \sum_{i=1}^{3} \alpha_i \xi^{(i,k)}$ with $\alpha_i \geq 0$ and $\sum_{i=1}^{3} \alpha_i = 1$. Then we have

$$\int_\Pi f(y, \xi + \nabla_y N(\xi, y))dy \leq \int_\Pi \sum_i a_i f(y, \xi^{(i,k)} + \nabla_y v^{(i,k)}(y))dy$$

$$\leq \sum_i a_i f_{hom}(\xi^{(i,k)}) + \frac{\delta}{2|\Omega|}$$

$$\leq f_{hom}(\xi) + \frac{\delta}{2|\Omega|} + 3C_3 d_T \leq f_{hom}(\xi) + \frac{\delta}{|\Omega|}.$$

For $u_0(x) \in C^\infty_c(\Omega)$, define the recovery sequence as

$$u_\varepsilon(x) = u_0(x) + \varepsilon N(u_0(x), \frac{x}{\varepsilon}). \tag{10.52}$$

For $x \in \partial\Omega$, since $u_0(x) = 0$ then $N(u_0(x), \frac{x}{\varepsilon}) = v^{(0,0)}(\frac{x}{\varepsilon}) = 0$ and we conclude that $u_\varepsilon \in W^{1,1}_0(\Omega)$. By construction of N we have $u_\varepsilon \to u_0$ in $L^1(\Omega)$, and moreover

$$\int_\Omega f(\frac{x}{\varepsilon}, \nabla_x u_\varepsilon(x)) dx = \int_\Omega f(\frac{x}{\varepsilon}, \nabla_x u_0(x) + \varepsilon \nabla_x N(u_0(x), \frac{x}{\varepsilon})) dx$$

$$\leq \int_\Omega f(\frac{x}{\varepsilon}, \nabla_x u_0(x) + \nabla_y N(u_0(x), \frac{x}{\varepsilon})) dx + O(\varepsilon). \tag{10.53}$$

Since $N(\xi, y)$ is continuous in ξ and $u_0(x)$ is continuous in x we have $\nabla_y N(u_0(x), y)$ is continuous in x. Using the generalized Averaging Lemma 6.1 and (10.53) we conclude

$$\limsup_{\varepsilon \to 0} \int_\Omega f(\frac{x}{\varepsilon}, \nabla_x u_\varepsilon(x)) \, dx \le \int_\Omega \int_\Pi f(y, \nabla_x u_0(x) + \nabla_y N(u_0(x), y)) \, dy \, dx$$

$$\le \int_\Omega f_{hom}(\nabla_x u_0(x)) \, dx + \delta. \tag{10.54}$$

$\square$

Proof *(Proof of Theorem 10.3)*
We first prove $\limsup_{\varepsilon \to 0} \tau_\varepsilon \le \tau_{hom}$. We have

$$\tau_{hom} = \inf_{\substack{u \in W_0^{1,1}(\Omega) \\ \int_\Omega pu\,dx=1}} \int_\Omega f_{hom}(\nabla u)\,dx. \tag{10.55}$$

By Lipschitz continuity of f_{hom}, for all $\delta > 0$, there exists $u_0(x) \in C_c^\infty(\Omega)$ such that

$$\int_\Omega pu_0\,dx = 1, \quad \int_\Omega f_{hom}(\nabla u_0)\,dx \le \tau_{hom} + \frac{\delta}{2}. \tag{10.56}$$

Using Lemma 10.5, we construct $\{u_\varepsilon\} \subset W_0^{1,1}(\Omega)$, $u_\varepsilon \to u_0$ in $L^1(\Omega)$ such that

$$\limsup_{\varepsilon \to 0} \int_\Omega f(\frac{x}{\varepsilon}, \nabla u_\varepsilon)\,dx \le \int_\Omega f_{hom}(\nabla u_0)\,dx + \frac{\delta}{2}. \tag{10.57}$$

Define $\lambda_\varepsilon := \int_\Omega pu_\varepsilon \, dx$. Since $\lambda_\varepsilon \to 1$, we have

$$\limsup_{\varepsilon \to 0} \tau_\varepsilon \le \limsup_{\varepsilon \to 0} \int_\Omega f(\frac{x}{\varepsilon}, \lambda_\varepsilon^{-1} \nabla u_\varepsilon) \, dx = \int_\Omega f_{hom}(\nabla u_0) \, dx + \frac{\delta}{2} \le \tau_{hom} + \delta. \tag{10.58}$$

As δ is arbitrary we have proved $\limsup_{\varepsilon \to 0} \tau_\varepsilon \le \tau_{hom}$.

We next show that $\liminf_{\varepsilon \to 0} \tau^\varepsilon \ge \tau_{hom}$. By Lemma 10.2 it suffices to prove the result for $p \in C^\infty(\Omega)$, as $C^\infty(\Omega)$ is a dense subset of $L^n(\Omega)$. Let $v_\varepsilon(x) = v_0(x) + Z(\frac{x}{\varepsilon}, v_0(x))$ be the vector field constructed in Lemma 10.4. Note that $Z(\frac{x}{\varepsilon}, v_0(x))$ is not divergence free but div $Z(\frac{x}{\varepsilon}, v_0(x)) \rightharpoonup 0$, so we add a small correction term to get a divergence free test function. Namely consider

$$\hat{v}_\varepsilon = v_\varepsilon + \nabla w_\varepsilon \tag{10.59}$$

where w_ε solves

$$-\Delta w_\varepsilon = \gamma(x, \frac{x}{\varepsilon}), \tag{10.60}$$

with Dirichlet boundary conditions $w_\varepsilon|_{\partial\Omega} = 0$. Since $\gamma(x, \frac{x}{\varepsilon}) \rightharpoonup 0$ in L^β for all $\beta > 1$, it follows from classical elliptic estimates for the Laplace operator [52] that

$$\limsup_{\varepsilon \to 0} |\nabla w_\varepsilon| = 0. \tag{10.61}$$

Then by convexity of g, for all $\delta > 0$ we have

$$\limsup_{\varepsilon \to 0} \int_\Omega g(\frac{x}{\varepsilon}, (1 - \delta)\hat{v}_\varepsilon)dx \leq (1 - \delta) \limsup_{\varepsilon \to 0} \int_\Omega g(\frac{x}{\varepsilon}, v_\varepsilon)dx$$
$$+ \delta \limsup_{\varepsilon \to 0} \int_\Omega g(\frac{x}{\varepsilon}, \frac{1 - \delta}{\delta}\nabla w_\varepsilon)dx. \tag{10.62}$$

The first term on the right-hand side of (10.62) is 0 by construction of v_ε. Also the domain of $g(y, \eta)$ contains a neighborhood of 0 for every y so that (10.61) implies

$$\int_\Omega g(\frac{x}{\varepsilon}, \frac{1 - \delta}{\delta}\nabla w_\varepsilon)dx = 0 \tag{10.63}$$

for sufficiently small ε. Thus we have

$$\liminf_{\varepsilon \to 0} \tau_\varepsilon((1 - \delta)p) \geq \tau_{hom}(p) \tag{10.64}$$

for all $\delta > 0$. Finally we use Lemma 10.2 to conclude the proof of Theorem 10.3.   □

Continuum Limits for Discrete Problems with Fine Scales

© Springer Nature Switzerland AG 2018
L. Berlyand, V. Rybalko, *Getting Acquainted with Homogenization and Multiscale,*
Compact Textbooks in Mathematics,
https://doi.org/10.1007/978-3-030-01777-4_11

A physical system can be modeled on different scales or resolutions. For example, a model of a fluid or gas can be considered at the levels of

1. the macroscopic scale with continuum mechanics models such as the Navier-Stokes or Euler equations. These models are often called phenomenological.
2. the molecular or atomistic scale where the particles move according to Newton's laws and are coupled with interacting potentials (e.g., molecular dynamics method).
3. the quantum scale with models such as the Schrödinger equation.

Obviously models on these levels are related because they describe the same physical systems. The challenging task is to justify the phenomenological models at the macroscopic scale by deriving them from atomistic models. The link between the continuum models and the atomistic nature of the matter was a long-time fundamental question in physics and mechanics and only recently was addressed in a rigorous mathematical context.

In this section we consider 1D "toy" examples of transition from atomistic to continuum models. Continuum approximations are very useful and quite often indispensable in the quantitative analysis of discrete systems with a very large number of degrees of freedom. These continuum approximations are obtained in this section by means of Γ-convergence for simple 1D models of a large number of material points interacting through some conservative forces. In such discrete problems the microscale enters as the lattice spacing Δ_n representing, e.g., interatomic distances. Then the continuum approximation is obtained by taking Γ-limit as $\Delta_n \to 0$. This approximation is a macroscopic (homogenized) model which contains no microscales. The key difference between the continuum two-scale problems that we studied previously and discrete problems is that the discrete problems contain the "range of interaction parameter" K. This parameter represents the number of neighbors that interact with a given material point. We will see that the homogenized macroscopic problem essentially depends on the value of this parameter. The other crucial feature of discrete problems is the pairwise interaction energies (potentials) that are typically nonlinear, which is why the Γ-convergence approach is useful.

In this section we follow Chapter 4 of [28] with significantly more details of proofs and heuristic explanations supplied by figures and devise a new counterexample illustrating the failure of periodic convexification formula for long range interactions (Example 11.2). We also would like to acknowledge that our understanding of these problems benefited from reading other papers by A. Braides and his webpage.

11.1 Discrete Spring-Particle Problems in 1D and Homogenization for Convex Energies

On the interval $[0, 1]$ we consider the set of $n + 1$ equidistant points $x_i = i/n$, $i = 0, 1, \ldots, n$ which serves as the reference configuration of material points. Let u_i be the displacement of i-th point. We assume that the energy of interaction of i-th and j-th points $(j > i)$ is

$$\phi_l \left(\frac{u_j - u_i}{x_j - x_i} \right), \quad \text{where } \phi_l(s) \text{ is a given function, } l = j - i. \tag{11.1}$$

Summing all of these pairwise energies leads to the total energy $E_n(\{u_i\})$, which, after a renormalization by the factor $\Delta_n := 1/n$, takes the form

$$E_n(\{u_i\}) = \sum_{l=1}^{K} \sum_{i=0}^{n-l} \Delta_n \phi_l \left(\frac{u_{i+l} - u_i}{l \Delta_n} \right). \tag{11.2}$$

Here the above renormalization makes the sum of $O(n)$ terms in (11.2) finite. The number K in (11.2) represents the maximal range of interaction, i.e., for instance, if $K = 1$, then we have only nearest neighbor interactions, if $K = 2$, then next-to-nearest neighbor interactions are additionally present, etc.

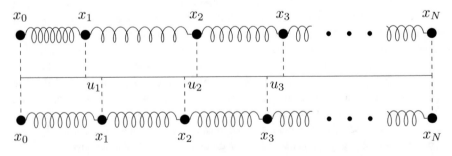

⬛ Fig. 11.1 1D system of particles connected via springs with nearest neighbor interactions

141

11

11.1 · Discrete Spring-Particle Problems in 1D and Homogenization for...

Note that (11.2) is energy of a discrete system that contains two scales: the macroscopic size of the interval $[0, 1]$ and the microscopic size of the lattice spacing $\Delta_n \to 0$ as $n \to \infty$. Our ultimate goal is to obtain a continuum approximation for the following minimization problem:

$$\min_{\{u_i\}} \left\{ E_n(\{u_i\}) - \sum_{i=1}^{n} u_i f_i \Delta_n \right\} \quad \text{subject to some boundary conditions, e.g. } u_0 = u_n = 0,$$

(11.3)

in the limit of a large number n of material points. Here f_i are (given) external forces applied to the points. The simplest model problem in this framework is that of n point masses connected to their nearest neighbors by identical linear springs, which corresponds to the case of nearest neighbor interactions $K = 1$ and quadratic energies

$$\phi_1(s) = \frac{1}{2}\kappa s^2,$$

(11.4)

κ being the stiffness of the springs, see ◻ Figure 11.1. As one can expect, in this simple case problem (11.3) leads in the limit $n \to \infty$ to the continuous minimization problem

$$\min_u \left\{ \int_0^1 \kappa(u')^2 - uf \, dx \right\}$$

with some boundary conditions. The latter problem models horizontal displacement of a homogeneous spring subject to the load $f(x)$. This example is quite simple due to the following two reasons:

- The energy (11.4) is quadratic (linear spring) which corresponds to a linear Euler-Lagrange equation or linear Hooke's law. In fact, what matters is that the energy (11.4) is *convex* as homogenization of nonconvex problems is a harder task. This phenomenon has already been seen in the homogenization of continuum problems in ▶ Section 9.

- Only nearest neighbors interact, that is $K = 1$ in (11.2). To explain the role of K recall the notion of *non-local* differential equations. Standard local differential equations relate the values of the unknown function and its derivatives at a given point. By contrast, a non-local differential equation relates the value of the unknown function at a given point to the values of its derivatives or the function itself at other points in the domain. A simple example is an integro-differential equation $u'(x) = \int_D F(x, y)u(y)dy$, where the integral is taken over a subdomain D whose size characterizes the range of interactions in the system. In discrete problems this range of interaction is determined by the parameter K and non-locality means that $K > 1$. Actually, for fixed K the range of interaction $K\Delta_n$ is still small and the limiting problem is local. However for $K > 2$ the formula for homogenized Lagrangian changes significantly as will be shown in ▶ Section 11.4. Roughly speaking for $K = 1, 2$ the homogenized Lagrangian is determined by a standard cell problem with period 1 or 2, respectively, whereas for $K > 2$ an expanding cell problem must be introduced. This can be compared to expanding cell problem that appears in the homogenization of nonconvex continuum problems ($\blacktriangleright$ Section 9).

Note that discrete variational problems (11.3) are defined on discrete functions (n tuples) that depend on n. Therefore, in order to define a notion of convergence, all such functions have to be embedded into a single space. This is done by passing from tuples to associated piecewise linear functions. That is discrete tuples $\{u_i\}$ are identified with their continuous, piecewise linear interpolations $u(x)$:

$$u(x) := u_i + (u_{i+1} - u_i)\frac{x - x_i}{\Delta_n}, \quad x \in [x_i, x_{i+1}], i = 0, 1, \ldots, (n-1). \quad (11.5)$$

Hereafter the class of all piecewise linear functions $u(x)$ of the form (11.5) will be denoted by $\mathscr{L}_n$. Then

$$\frac{u_{i+1} - u_i}{\Delta_n} = u'(x) \text{ and } \frac{u_{i+l} - u_i}{l\Delta_n} = \frac{1}{l\Delta_n}\sum_{m=i}^{i+l-1}(u_{m+1} - u_m) = \frac{1}{l}\sum_{m=0}^{l-1}u'(x + m\Delta_n),$$

when $x \in (x_i, x_{i+1})$, so

$$\Delta_n\phi_l\left(\frac{u_{i+l} - u_i}{l\Delta_n}\right) = \int_{x_i}^{x_{i+1}}\phi_l\left(\frac{u_{i+l} - u_i}{l\Delta_n}\right)dx = \int_{x_i}^{x_{i+1}}\phi_l\left(\frac{1}{l}\sum_{m=0}^{l-1}u'(x + m\Delta_n)\right)dx.$$

Thus we can write the energy functional $E_n(\{u_i\})$ in the integral form

$$E_n(\{u_i\}) = E_n^{(c)}(u), \quad (11.6)$$

where

$$E_n^{(c)}(u) := \sum_{l=1}^{K}\int_0^{1-\Delta_n(l-1)}\phi_l\left(\frac{1}{l}\sum_{m=0}^{l-1}u'(x + m\Delta_n)\right)dx. \quad (11.7)$$

Here the upper integration limit $1 - \Delta_n(l-1)$ is chosen in accordance with (11.2). Finally, we extend the functional $E_n^{(c)}(u)$ to the entire Sobolev space $W^{1,1}(0, 1)$ by setting $E_n^{(c)}(u) = +\infty$ if $u \notin \mathscr{L}_n$ (that is we exclude all but piecewise linear functions) to reformulate the minimization problem (11.3) in the form

$$\min_{u\in W^{1,1}(0,1),\, u(0)=u(1)=0}\left\{E_n^{(c)}(u) - \sum_{i=1}^{n}u(x_i)f_i\Delta_n\right\}. \quad (11.8)$$

Note that this extension does not affect minimization since functions $u \notin \mathscr{L}_n$ have infinite energy and therefore cannot be minimizers. The growth of the interaction energies $\phi_l(s)$ as $|s| \to \infty$ plays an important role. The space $W^{1,1}(0, 1)$ is chosen

11.1 · Discrete Spring-Particle Problems in 1D and Homogenization for...

143

11

because the growth of the interaction energies is not specified yet (see (11.9)) and $W^{1,1}(0, 1)$ contains $W^{1,p}(0, 1)$, $p > 1$.

Since we are interested in the asymptotic behavior of minimization problems (11.3) or equivalently (11.8), it is natural to employ Γ-convergence to compute a continuum (homogenization) limit of the discrete energies $E_n(\{u_i\})$ as $n \to \infty$. That is, we seek a Γ-limit of the functionals $E_n^{(c)}$, which (if it exists) is typically an integral functional of the form $E(u) = \int_0^1 L(u')dx$, where the homogenized Lagrangian $L(s)$ defines the energy density of the limiting continuum macroscopic system. The simplest case is that of nearest neighbor interactions with positive convex energies. That is assume that the interaction range $K = 1$ and $\phi_1(s)$ is a convex function with a superlinear growth at ∞ ($|\phi_1(s)|/|s| \to +\infty$ as $|s| \to \infty$). Then the functionals $E_n^{(c)}$ Γ-converge to the integral functional $E(u) = \int_0^1 \phi_1(u')dx$ with respect to the weak convergence in $W^{1,1}(0, 1)$. Thus taking the Γ-limit in this case simply amounts to replacing the difference quotients $\frac{u_{i+1} - u_i}{\Delta_n}$ by the derivative $u'(s)$. This result is a particular case of the Γ-convergence result for longer range interactions $K \geq 1$ with convex energies which will be stated and proved under growth conditions that are a bit more restrictive than the superlinear growth condition. Specifically, assume that $K \geq 1$ is fixed and $\forall l = 1 \ldots K$

$$C_1|s|^p \leq \phi_l(s) \leq C_2(|s|^p + 1) \text{ with } 0 < C_1 \leq C_2 \text{ and } p > 1. \tag{11.9}$$

Note that some growth condition is necessary for the existence of the minimizer (cf. Definition 9.2 and condition (9.6)) and we choose $p > 1$ in order to work in the Banach space $W^{1,p}(0, 1)$, which is reflexive unlike $W^{1,1}(0, 1)$ (recall that bounded sequences in reflexive spaces are weakly compact and this property is necessary for establishing for $\Gamma-$ convergence). As a specific example of energies one can consider $\phi_l(s) = |s|^p + |s|^{p_l}$, $1 \leq p_l \leq p$.

Theorem 11.1

For a given integer $K \geq 1$, assume that $\phi_l(s)$, $l = 1, \ldots K$ are convex functions satisfying the condition (11.9). Then functionals $E_n^{(c)}$ Γ-converge in $L^p(0, 1)$ to the functional

$$E(u) = \begin{cases} \int_0^1 \sum_{l=1}^K \phi_l(u')dx \text{ if } u \in W^{1,p}(0, 1) \\ +\infty, \text{ otherwise.} \end{cases} \tag{11.10}$$

ⓘ **Remark 11.1** It follows from (11.10) that the homogenized Lagrangian for convex energies is the same as the original interaction energy. We will see below that it is no longer the case for nonconvex energies.

Proof. Thanks to (11.9), it is sufficient to establish Γ-convergence with respect to weak $W^{1,p}(0, 1)$ convergence instead of strong L^p convergence. This can be shown using the argument from part A in the proof of Theorem 9.3. Since $E_n^{(c)}(u_n) = +\infty$ if $u_n \notin \mathscr{L}\backslash$ it is sufficient to consider we consider only piecewise piecewise linear functions $u_n(x)$, $u_n \in \mathscr{L}_n$ that weakly converge in $W^{1,p}(0, 1)$, cf. step A in the proof of Theorem 9.3. Let $u(x)$ be the weak limit, i.e., $u_n \rightharpoonup u$ in $W^{1,p}(0, 1)$. Then the functions

$$v_{l,n}(x) := \frac{1}{l} \sum_{m=0}^{l-1} u_n(x + m \Delta_n) \tag{11.11}$$

appearing in energy functional (11.7), converge to $u(x)$ as $n \to \infty$ weakly in $W^{1,p}(0, 1 - \delta)$ for every $\delta > 0$. This follows from definition of weak convergence and the fact that $\Delta_n \to 0$ as $n \to \infty$. Indeed, when n is sufficiently large it holds that

$$\int_0^{1-\delta} u_n(x + m\Delta_n)\varphi(x)dx = \int_0^{1-\delta} u_n(y)\varphi(y - m\Delta_n)dy$$

for any continuous test function φ with compact support in $(0, 1-\delta)$, and $\varphi(y - m\Delta_n) \to \varphi(y)$ uniformly as $n \to \infty$. Fix $\delta > 0$ so that $\delta > K\Delta_n$ in (11.11). Thus

$$\liminf_{n \to \infty} E_n^{(c)}(u_n) \geq \liminf_{n \to \infty} \sum_{l=1}^{K} \int_0^{1-\delta} \phi_l\left(v_{l,n}'(x)\right) dx \geq \sum_{l=1}^{K} \int_0^{1-\delta} \phi_l\left(u'(x)\right) dx.$$

$$\tag{11.12}$$

Here the first inequality is due to (11.7), the choice of δ and positivity $\phi_l(s) \geq 0$, while in the second inequality we have used the fact that the integral functionals are weakly lower semicontinuous, thanks to convexity of the Lagrangians $\phi_l(s)$. Letting $\delta \to 0$ in the right-hand side of (11.12) we establish the required $\Gamma - \liminf$ inequality.

To show the $\Gamma - \limsup$ inequality, consider $u \in W^{1,p}(0, 1)$ and construct a recovery sequence by choosing $u_n(x)$ to be the piecewise linear interpolation of u at points x_i. Observe that

$$\frac{1}{l} \sum_{m=0}^{l-1} u_n'(x + m\Delta_n) = \frac{1}{x_{i+1} - x_i} \int_{x_i}^{x_{i+l}} u'(t)\, dt \quad \text{for } x \in [x_i, x_{i+1}]. \tag{11.13}$$

Then

$$\int_{x_i}^{x_{i+1}} \phi_l\left(\frac{1}{l} \sum_{m=0}^{l-1} u_n'(x + m\Delta_n)\right) dx = \Delta_n \phi_l\left(\frac{1}{l} \sum_{m=0}^{l-1} u_n'(x + m\Delta_n)\right)$$

$$\leq \frac{1}{l} \int_{x_i}^{x_{i+1}} \phi_l\left(u'(x)\right) dx. \tag{11.14}$$

Here the first equality is obtained from (11.13) whose left-hand side does not depend on x (on $[x_i, x_{i+1}]$) while the upper bound is obtained by using (11.13) in the left-hand side of the inequality followed by applying Jensen's inequality.

Thus, in view of (11.14),

$$
E_n^{(c)}(u_n) = \sum_{l=1}^{K} \int_0^{1-\Delta_n(l-1)} \phi_l\left(\frac{1}{l}\sum_{m=0}^{l-1} u_n'(x + m\Delta_n)\right) dx
$$

$$
\leq \sum_{l=1}^{K} \frac{1}{l} \sum_{i=0}^{n-l} \int_{x_i}^{x_{i+l}} \phi_l\left(u'(x)\right) dx \leq \sum_{l=1}^{K} \int_0^1 \phi_l\left(u'(x)\right) dx,
$$

and the theorem is proved.

11.2 Nonconvex Problems with Nearest Neighbor Interactions

We now turn to the case of nonconvex energies $\phi_l(s)$, when both the result and the techniques are more interesting and difficult. First we consider only the nearest neighbor interactions $K = 1$. Here as above in (11.6)–(11.7) we represent the discrete energies by integral functionals of the form:

$$
E_n^{(c)}(u) := \begin{cases} \int_0^1 \phi\left(u'(x)\right) dx & \text{if } u \in \mathscr{L}_n \\ +\infty & \text{otherwise} \end{cases} \tag{11.15}
$$

(note that $E_n^{(c)}(u)$ depends on n only through its domain $\mathscr{L}_n$). The Lagrangian $\phi(s)$ in (11.15) is a continuous function, *not necessarily convex*, satisfying

$$
C_1|s|^p \leq \phi(s) \leq C_2(|s|^p + 1) \text{ with } 0 < C_1 \leq C_2 \text{ and } p > 1, \tag{11.16}
$$

Under these conditions, the $\Gamma-$ limit as $n \to \infty$ is the energy functional whose energy density ϕ^{**} is the convex envelope of ϕ. In other words, we have here the effect of convexification when taking the Γ-limit as $n \to \infty$. Recall that the convex envelope ϕ^{**} of a function ϕ is defined by

$$
\phi^{**}(s) := \sup\{\varphi(s) : \varphi \text{ is convex and } \varphi(x) \leq \phi(x) \ \forall x \in \mathbb{R}\}, \tag{11.17}
$$

i.e., in order to define $\phi^{**}(s)$ one first considers convex functions $\varphi(x)$ whose graphs lie under $\phi(x)$ for *all* x and then obtains the value of ϕ^{**} at s maximizing $\phi(s)$ among all such functions $\phi(x)$, see ◻ Figure 11.2.

Exercise 11.1

Let ϕ satisfy (11.16), prove that its convex envelope ϕ^{**} is obtained by applying the Legendre transform $f^*(p) = \sup_{s \in \mathbb{R}}(ps - f(s))$ twice, i.e., $\phi^{**} = (\phi^*)^*$. Hint: $f = (f^*)^*$ for convex functions f on $\mathbb{R}$ growing superlinearly ($\lim_{|x| \to \infty} |f(x)|/|x| = +\infty$).

∎

ⓘ Remark 11.2 Note that the convex envelope ϕ^{**} does not necessarily coincide with ϕ on intervals where ϕ is convex, see ▢ Figure 11.2 (ϕ is convex in a neighborhood of α_+ but $\phi^{**}(\alpha) = \phi(\alpha)$ only for $\alpha \geq \alpha_+$).

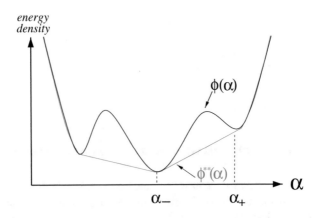

▢ Fig. 11.2 The curve in red is the convex envelope $\phi^{**}(\alpha)$ of $\phi(\alpha)$, α is the derivative (slope).

The following proposition provides yet another description of convex envelopes ϕ^{**} in terms of an auxiliary minimization problem. We will rely on this proposition in the proof of Γ-convergence for nonconvex energies in Theorem 11.2.

Proposition 11.1 *Assume that ϕ is a continuous function (not necessarily convex) satisfying growth condition (11.16). For given $\alpha \in \mathbb{R}$ consider the following minimization problem with respect to unknowns α_-, α_+, and λ:*

$$m(\alpha) = \inf\{\lambda\phi(\alpha_-) + (1 - \lambda)\phi(\alpha_+)|\, \lambda\alpha_- + (1 - \lambda)\alpha_+ = \alpha,$$

$$\alpha_- \leq \alpha \leq \alpha_+,\ 0 \leq \lambda \leq 1\}. \tag{11.18}$$

Then $\phi^{**}(\alpha) = m(\alpha)$. Moreover, if $\phi(\alpha) = \phi^{**}(\alpha)$, then there exists a minimizing triple $(\alpha_-, \alpha_+, \lambda)$ such that $\alpha_+ = \alpha_- = \alpha$ and $\lambda \in [0, 1]$ is arbitrary; if $\phi(\alpha) > \phi^{**}(\alpha)$, then strict inequalities $\alpha_- < \alpha < \alpha_+, 0 < \lambda < 1$ for a minimizer triple $(\alpha_-, \alpha_+, \lambda)$ hold.

ⓘ Remark 11.3 The idea of the proof is illustrated by ▢ Figure 11.3, where the straight dashed line moves downward until it becomes tangent to the graph of ϕ at the actual minimizers α_+ and α_- (i.e., α_+ and α_- for which the minimum in (11.18) is attained). The last statement of the Proposition 11.1 is illustrated by the middle portion in ▢ Figure 11.3 where red and black curves do not coincide.

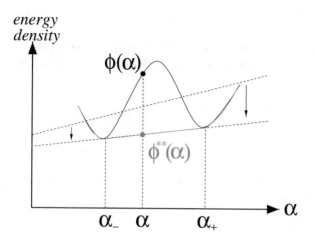

□ Fig. 11.3 Convexification via minimization problem (11.18).

Proof

Fix $\alpha \in \mathbb{R}$ and consider two cases $\phi(\alpha) = \phi^{**}(\alpha)$ and $\phi(\alpha) > \phi^{**}(\alpha)$ (from the definition of convex envelope it follows that $\phi(\alpha) \geq \phi^{**}(\alpha)$).

Case I: $\phi(\alpha) = \phi^{**}(\alpha)$.

Consider the auxiliary minimization problem

$$\overline{m}(\alpha) = \inf\{\lambda\phi^{**}(\alpha_-) + (1-\lambda)\phi^{**}(\alpha_+) \mid \lambda\alpha_- + (1-\lambda)\alpha_+ = \alpha,$$

$$\alpha_- \leq \alpha \leq \alpha_+, \ 0 \leq \lambda \leq 1\}. \tag{11.19}$$

Since $\phi^{**}(\alpha) \leq \phi(\alpha)$ for all $\alpha \in \mathbb{R}$, we have $m(\alpha) \geq \overline{m}(\alpha)$. On the other hand, $\overline{m}(\alpha) = \phi^{**}(\alpha)$ due to the convexity of $\phi^{**}(\alpha)$ and definition of $\overline{m}(\alpha)$ in (11.19) via minimization. Taking $\alpha_- := \alpha$ and $\lambda := 1$ in (11.18) we also get $m(\alpha) \leq \phi(\alpha)$. Thus

$$\phi^{**}(\alpha) = \overline{m}(\alpha) \leq m(\alpha) \leq \phi(\alpha), \tag{11.20}$$

and therefore $m(\alpha) = \phi^{**}(\alpha)$.

Case II: $\phi(\alpha) > \phi^{**}(\alpha)$ (middle of portion on □ Figure 11.3).

Step 1. Here we claim that ϕ^{**} is linear in a neighborhood of α. To this end we introduce the linear interpolation of ϕ^{**} on a sufficiently small interval $[\overline{\alpha}_-, \overline{\alpha}_+]$,

$$f_{\overline{\alpha}_-, \overline{\alpha}_+}(\theta) := \phi^{**}(\overline{\alpha}_-)\frac{\theta - \overline{\alpha}_+}{\overline{\alpha}_- - \overline{\alpha}_+} + \phi^{**}(\overline{\alpha}_+)\frac{\theta - \overline{\alpha}_-}{\overline{\alpha}_+ - \overline{\alpha}_-}, \tag{11.21}$$

and show that

$$\phi^{**}(\theta) = f_{\overline{\alpha}_-,\overline{\alpha}_+}(\theta) \quad \text{when } \theta \in [\overline{\alpha}_-, \overline{\alpha}_+].$$ (11.22)

Note that $f_{\overline{\alpha}_-,\overline{\alpha}_+}(\theta)$ is actually a two-parameter family of linear functions.

Since ϕ^{**} is continuous (being convex and taking finite values), it follows from (11.21) that

$$\max_{[\overline{\alpha}_-,\overline{\alpha}_+]} f_{\overline{\alpha}_-,\overline{\alpha}_+}(\theta) \to \phi^{**}(\alpha) \quad \text{when } \overline{\alpha}_+ \to \alpha \text{ (from the right) and } \overline{\alpha}_- \to \alpha \text{ (from the left).}$$ (11.23)

Since $\phi^{**}(\alpha) < \phi(\alpha)$ and due to continuity of ϕ we choose $\overline{\alpha}_- < \alpha$ and $\overline{\alpha}_+ > \alpha$ sufficiently close to α such that

$$f_{\overline{\alpha}_-,\overline{\alpha}_+}(\theta) \leq \phi(\theta)$$ (11.24)

for all θ in $[\overline{\alpha}_-, \overline{\alpha}_+]$. Next extend (11.24) to $\mathbb{R}$. Indeed, due to convexity of ϕ^{**} we have $f_{\overline{\alpha}_-,\overline{\alpha}_+}(\theta) \leq \phi^{**}(\theta)$ for $\theta \notin [\overline{\alpha}_-, \overline{\alpha}_+]$ (i.e., the graph of the linear function $f_{\overline{\alpha}_-,\overline{\alpha}_+}$ intersects that of ϕ^{**} at $\overline{\alpha}_-$ and $\overline{\alpha}_+$ and due to convexity of ϕ^{**} we conclude $\phi^{**}(\theta) \geq f_{\overline{\alpha}_-,\overline{\alpha}_+}(\theta)$ for all $\theta \notin [\overline{\alpha}_-, \overline{\alpha}_+]$). Since $\phi^{**}(\theta) \leq \phi(\theta)$, we have that (11.24) also holds for $\theta \notin [\overline{\alpha}_-, \overline{\alpha}_+]$.

Next, observe that $f_{\overline{\alpha}_-,\overline{\alpha}_+}$ is a convex (linear) and therefore the definition of the convex envelope ϕ^{**} implies

$$f_{\overline{\alpha}_-,\overline{\alpha}_+}(\theta) \leq \phi^{**}(\theta) \quad \forall \theta \in \mathbb{R}.$$ (11.25)

Combining (11.24), (11.25) and taking into account that $\phi^{**}(\theta) \leq f_{\overline{\alpha}_-,\overline{\alpha}_+}(\theta)$ (see ◻ Figure 11.3), we arrive at

$$f_{\overline{\alpha}_-,\overline{\alpha}_+}(\theta) = \phi^{**}(\theta) \quad \text{for } \theta \in [\overline{\alpha}_-, \overline{\alpha}_+].$$ (11.26)

Thus, the claim of Step 1 is proved.

Step 2. Consider now the maximal closed interval $[\alpha_-, \alpha_+]$ of linearity of $\phi^{**}(\theta)$. The interval $[\alpha_-, \alpha_+]$ is finite due to coercivity of ϕ, which ensures that $\phi^{**}(\theta)$ grows as $|\theta|^p$ when $|\theta| \to \infty$. The maximality of $[\alpha_-, \alpha_+]$ implies that

$$\phi^{**}(\alpha_\pm) = \phi(\alpha_\pm).$$ (11.27)

This can be seen from ◻ Figure 11.3, for the rigorous argument assume $\phi^{**}(\alpha_+) < \phi(\alpha_+)$, then one can repeat arguments of the Step 1 with α_+ in place of α and thus extend the interval of linearity of ϕ^{**} to the right from α_+. This contradicts the maximality of $[\alpha_-, \alpha_+]$ and therefore (11.27) holds.

Now substitute these endpoints $\alpha_\pm$ into (11.18) and setting $\lambda := (\alpha_+ - \alpha)/(\alpha_+ - \alpha_-)$ we get

$$m(\alpha) \leq \lambda\phi(\alpha_-) + (1-\lambda)\phi(\alpha_+) = \lambda\phi^{**}(\alpha_-) + (1-\lambda)\phi^{**}(\alpha_+) = \phi^{**}(\alpha).$$ (11.28)

Here we used (11.27) and the last equality follows from linearity of ϕ^{**} in the interval $[\alpha_-, \alpha_+]$. Finally observe that (11.20) holds in the Case II ($\phi(\alpha) > \phi^{**}(\alpha)$) by exactly the same argument as in the Case I ($\phi(\alpha) = \phi^{**}(\alpha)$). In particular we have $m(\alpha) \geq \phi^{**}(\alpha)$, therefore (11.28) implies that $m(\alpha) = \phi^{**}(\alpha)$. $\qquad\qquad\square$

Theorem 11.2

Let ϕ be a continuous function satisfying (11.16). Then functionals $E_n^{(c)}$ corresponding to the nearest neighbor interactions (given by (11.15)) Γ-converge in $L^p(0, 1)$ to the functional

$$E(u) = \begin{cases} \int_0^1 \phi^{**}(u')dx \text{ if } u \in W^{1,p}(0, 1) \\ +\infty, \text{ otherwise.} \end{cases} \tag{11.29}$$

ⓘ **Remark 11.4** Theorem 11.2 describes the effect of convexification in homogenization of discrete problems that is the homogenized Lagrangian in (11.29) is the convex envelope of the original Lagrangian in (11.15). The same effect is present in continuum problems. Indeed, consider the Γ-limit of a sequence that consists of identical functionals $E_n(u) = E(u) := \int_0^1 L(u')dx$ with nonconvex Lagrangian L. Assume that $L(s)$ is continuous and grows superlinearly at ∞. Then the Γ-limit with respect to the weak convergence in $W^{1,1}(0, 1)$ is $E^{**}(u) = \int_0^1 L^{**}(u')dx$. The proof of this fact is similar to the proof of Theorem 11.2. In contrast, as noted at the beginning of this section, the Γ-limits of discrete problems with a larger range of interactions $K > 2$ are quite different from those of continuum problems. In particular, for discrete problems with $K > 2$ one must solve an expanding cell problem analogous to (9.29) which is no longer equivalent to a periodic cell problem.

Example 11.1

(A bistable spring system with nearest neighbor interactions, $K = 1$.)

Consider a model of material points connected by springs whose interactions are described by the following Lagrangian:

$$\phi_1(\alpha) = (|\alpha| - \beta)^2, \tag{11.30}$$

where β is a positive parameter describing equilibrium states of each spring which are assumed to be identical. Specifically, a spring has two equilibrium states: $\alpha = \beta$ (expanded state) and $\alpha = -\beta$ (contracted state), where $\alpha = (u_{i+1} - u_i)/\Delta_n$ is elongation of the spring measured in reference configuration length $\Delta_n = 1/n$. Here $\{u_i\}$ denotes displacements of the points from reference configuration $\{x_i = i/n\}$. Note that the reference configuration here is not an equilibrium state as in (11.4) unless $\beta = 0$ that corresponds to standard quadratic (harmonic) springs.

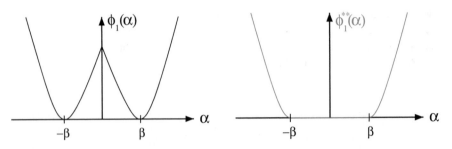

Fig. 11.4 Left: Interaction potential $\phi_1(\alpha)$. Right: Convex envelope $\phi^{**}(\alpha)$.

The potential (11.30) is depicted in ■ Figure 11.4 (left). Note that it consists of two pieces that correspond to contraction, $\alpha\Delta_n = u_{i+1} - u_i < 0$, and expansion, $\alpha\Delta_n = u_{i+1} - u_i > 0$. Clearly $\phi_1(\alpha)$ is nonconvex. By Theorem 11.2, the homogenized Lagrangian in this case is the convex envelop of $\phi_1(\alpha)$ which is given by

$$\phi_1^{**}(\alpha) = \begin{cases} (|\alpha| - \beta)^2 \text{ if } |\alpha| > \beta \\ 0 \text{ otherwise.} \end{cases}$$

Note that nonconvexity leads to "flat piece" of the homogenized Lagrangian (deformations occur without change of energy), see ■ Figure 11.4 (right). ■

ⓘ Remark 11.5 The key technical ingredient in the proof of this theorem relies on Proposition 11.1. Namely, the homogenized Lagrangian ϕ^{**} is obtained by convexification of the original Lagrangian ϕ which amounts to optimization between two slopes α_- and α_+ with proportions λ and $1 - \lambda$ in test functions (e.g., recovery sequences), see ■ Figure 11.5 and ■ Figure 11.6. The origin of the two slopes can be illustrated by Example 11.1 of the system composed of bistable springs in which two wells of the double well Lagrangian determine two slopes $\alpha_\pm = \pm\beta$.

Proof *(Proof of Theorem 11.2.)*
Since $\phi \geq \phi^{**}$, we have $E_n^{(c)}(u_n) \geq E(u_n)$. On the other hand, the functional $E(u)$ is lower semicontinuous with respect to the convergence in $L^p(0, 1)$, due to the fact that ϕ^{**} is convex and $C_1|s|^p \leq \phi^{**}(s) \leq C_2(|s|^p + 1)$. Here convexity provides weak lower semicontinuity of $E(u)$ in $W^{1,p}(0, 1)$ which in turn implies that $E(u)$ is lower semicontinuous with respect to strong convergence in $L^p(0, 1)$, thanks to the bound $\phi^{**}(s) \geq C_1|s|^p$ (cf. part A of the proof of Theorem 9.1 in section 9.4). Thus, if $u_n \to u$ in $L^p(0, 1)$, then $\liminf_{n\to\infty} E_n^{(c)}(u_n) \geq \liminf_{n\to\infty} E(u_n) \geq E(u)$.

Before verifying the second (lim sup) condition on $u \in W^{1,p}(0, 1)$ note that we can assume, without loss of generality, that u is a piecewise linear function on $[0, 1]$ (not to confuse with functions from $\mathscr{L}_n$ that are piecewise linear interpolations of discrete functions for a given fine partition x_i, $i = 1, ...n$). Indeed, such functions are dense in $W^{1,p}(0, 1)$ and $E(u)$ is continuous on $W^{1,p}(0, 1)$. If we can construct functions $u_{k,n} \in \mathscr{L}_n$ such that $u_{k,n} \to u_k$ in $L^p(0, 1)$ and $\limsup_{n\to\infty} E_n^{(c)}(u_{k,n}) = E(u_k)$ for every piecewise linear function u_k, then choosing $u_k \to u$ in $W^{1,p}(0,1)$ as $k \to \infty$, we can find $k(n) \to \infty$ such that $u_{k(n),n} \to u$ in $L^p(0,1)$ and $\limsup_{n \to \infty} E_n^{(c)}(u_{k,n}) \leq \lim_{k \to \infty} E(u_k) = E(u)$. Note that the continuity of $E(u)$ on $W^{1,p}(0,1)$ follows from the convexity of ϕ^{**} and the inequalities $C_1|s|^p \leq \phi^{**}(s) \leq C_2(|s|^p + 1)$ (cf. Exercise 9.6).

Next we construct the recovery sequence for piecewise linear u whose linearity intervals form a finite coarse partition $y_0 = 0, y_1, ..., y_M = 1$ of $[0,1]$, where M does not depend on n. For this function u, consider its arbitrary linearity interval $[a,b]$ with $a = y_k, b = y_{k+1}$ so that u on this interval is of the form

$$u(x) = \alpha(x - a) + u(a) \quad (u' = \alpha \text{ on } [a,b]). \tag{11.31}$$

Since a and b may not be in the set $\{x_i\}$, we choose from this set the two points nearest from the left to the endpoints a and b, $a(n) := \max_{i=0,...n}\{x_i \leq a\}$ and $b(n) = \max_{i=0,...n}\{x_i \leq b\}$. This "small" shift is needed since the recovery sequence u_n is in $\mathscr{L}_n$, that is it must be linear on each fine subinterval $[x_i, x_{i+1}]$.

We next describe two main properties (11.32) and (11.33) of the desired recovery sequence followed by actual construction.

We begin from the approximation property. We will construct piecewise linear functions $u_n(x)$ on course subintervals $[a(n), b(n)]$ and then "glue" them together to extend $u_n(x)$ to the entire interval $[0,1]$ and thus obtain a function from $\mathscr{L}_n$. Our construction on each coarse subinterval will satisfy the following approximation property and the boundary condition needed for the "gluing":

$$\max_{[a(n),b(n)]} |u_n(x) - u(x)| = O(\Delta_n) \quad \text{and} \quad u_n(a(n)) = u(a), \; u_n(b(n)) = u(b).$$

$$\tag{11.32}$$

Note carefully that the length of fine intervals is $\Delta_n = |x_i - x_{i-1}| = n^{-1}$ and $n \to \infty$, while the length of each coarse interval $|b(n) - a(n)|$ does not vanish as $n \to \infty$, and we aim to construct a recovery sequence u_n such that on each coarse interval $[a(n), b(n)]$ it satisfies (11.32). Repeating this construction on all intervals $[a(n), b(n)] = [y_k, y_{k+1}]$ of linearity of $u(x)$ (with the small shift of endpoints as described above), we obtain functions $u_n(x)$ such that $u_n \in \mathscr{L}_n$ and u_n uniformly converges to u on $[0,1]$, thanks to (11.32).

Next, we describe the energy convergence property for each $[a(n), b(n)]$:

$$\limsup_{n \to \infty} \int_{a(n)}^{b(n)} \phi\left(u_n'(x)\right) dx = \lim_{n \to \infty} \int_{a(n)}^{b(n)} \phi\left(u_n'(x)\right) dx = \int_a^b \phi^{**}\left(u'(x)\right) dx.$$

$$\tag{11.33}$$

Once this property is established, the desired lim sup bound follows. Note that due to linearity of u on $[a(n), b(n)]$ we have that $\phi^{**}\left(u'(x)\right) \equiv \phi^{**}(\alpha)$ in this interval and, thus,

$$\lim_{n \to \infty} \int_{a(n)}^{b(n)} \phi\left(u_n'(x)\right) dx = (b - a)\phi^{**}(\alpha). \tag{11.34}$$

After summing up over all the intervals of linearity of $u(x)$, this leads to the desired lim sup inequality (which is actually an equality), $\limsup_{n\to\infty} E_n^{(c)}(u_n) \le E(u)$ so that $u_n(x)$ is a recovery sequence on $(0, 1)$.

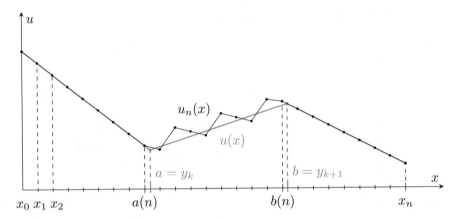

◻ Fig. 11.5 Construction of the recovery sequence.

We now proceed with the technical construction of functions $u_n(x)$ satisfying (11.32) and (11.33). This construction is particularly simple if $\phi^{**}(\alpha) = \phi(\alpha)$ (red and black curves coincide on ◻ Figure 11.3, where α is slope: $\alpha = u'(x)$). In this case we choose $u_n(x)$ to be the piecewise linear function such that $u_n(x_i) = u(x_i)$ if $x_i \in (a(n), b(n))$. In order to satisfy the second condition in (11.32) we choose $u_n(a(n)) = u(a)$, $u_n(b(n)) = u(b)$. It remains to observe that the slope of u_n is equal to α on each interval $[x_i, x_{i+1}] \subset (a(n), b(n))$ (see ◻ Figure 11.5, intervals where red and black curves coincide). This holds for all fine subintervals $[x_i, x_{i+1}]$ except possibly the left/rightmost intervals $[a(n), a(n) + \Delta_n]$ and $[b(n) - \Delta_n, b(n)]$, respectively, where the slope of u_n remains bounded as n increases. Since the length of these two intervals is bounded by $1/n$, the equality (11.33) follows.

Now suppose $\phi^{**}(\alpha) \ne \phi(\alpha)$ (middle portion in ◻ Figure 11.5). Then $\phi^{**}(\alpha) < \phi(\alpha)$ by the definition of ϕ^{**}. In this case we construct *oscillating* functions u_n with two slopes as shown in ◻ Figure 11.6. The slopes of these functions will take one of two values α_- and α_+ on each fine interval $[x_i, x_{i+1}]$ (possibly, except the last one), where α_+ and α_- solve the minimization problem (11.18) in Proposition 11.1. Specifically, u_n is defined by linear interpolation of the values $u_n(x_i)$ and these values are defined as follows. First we set $u_n(a(n)) := u(a)$ and then define successively

$$u_n(x_{i+1}) := \begin{cases} u_n(x_i) + \alpha_- \Delta_n & \text{if } u_n(x_i) \ge u(x_i), \\ u_n(x_i) + \alpha_+ \Delta_n & \text{if } u_n(x_i) < u(x_i). \end{cases}$$

Finally, at $x = b(n)$ we set $u_n(b(n)) = u(b)$, see ◻ Figure 11.6.

It follows from the construction of u_n in both cases $\phi^{**}(\alpha) = \phi(\alpha)$ and $\phi^{**}(\alpha) \ne \phi(\alpha)$ that

$$\max_{[a(n),b(n)]} |u_n(x) - u(x)| = O(\Delta_n). \tag{11.35}$$

Therefore we have

$$(N_-\alpha_- + N_+\alpha_+)\Delta_n = \alpha(b-a) + O(\Delta_n), \text{ and } (N_- + N_+)\Delta_n = (b-a) + O(\Delta_n),$$

(11.36)

where $N_\pm = N_\pm(n)$ denote the number of fine intervals (x_i, x_{i+1}) where $u'_n = \alpha_\pm$. Passing to the limit in (11.36) as $n \to \infty$ and taking into account the fact that $\lambda\alpha_- + (1-\lambda)\alpha_+ = \alpha$, we get

$$N_-(n)\Delta_n \to \lambda(b-a) \text{ and } N_+(n)\Delta_n \to (1-\lambda)(b-a).$$

(11.37)

That is, λ and $1 - \lambda$ are the proportions of intervals with slopes α_- and α_+, respectively.

Using (11.35), (11.37) and that $u'_n(x)$ is constant on each interval (x_i, x_{i+1}) (which has length Δ_n), we obtain

$$\int_{a(n)}^{b(n)} \phi\left(u'_n(x)\right) dx = \int_{a(n)}^{b(n)-\Delta_n} \phi\left(u'_n(x)\right) dx + O(\Delta_n)$$

$$= N_-(n)\Delta_n\phi(\alpha_-) + N_+(n)\Delta_n\phi(\alpha_+) + O(\Delta_n)$$

$$\to (b-a)(\lambda\phi(\alpha_-) + (1-\lambda)\phi(\alpha_+)) = (b-a)\phi^{**}(\alpha).$$

(11.38)

Thus, the energy convergence property (11.33) is established. Therefore, $\{u_n\}_n$ is indeed a recovery sequence. This completes the proof of the theorem. $\square$

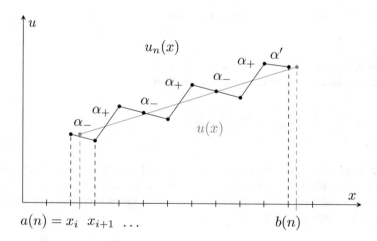

■ **Fig. 11.6** Oscillating recovery sequence $u_n(x)$

11.3 Nonconvex Problems with Long Range of Interactions

Consider now the case of nonconvex discrete energies (11.2) of a system of n points $\{x_i\}$ with long range interactions, $K \geq 2$, and Lagrangians $\phi_l(\alpha)$ satisfying the growth condition (11.9). The main result of this subsection is Theorem 11.4 where we establish the Γ-limit of these discrete energies and show that it is an integral functional with the Lagrangian $\overline{\phi}(\alpha)$ defined by means of expanding cell problems (11.59).

The heuristics for the derivation of the continuum Γ-limit can be presented in three main steps.

Step 1 (Introduction of a Mesoscale and a Partition of $\{x_i\}$)
We choose an integer R such that $n \gg R \gg K$ and assume for simplicity that n/R is also an integer. Then in addition to microscale $|x_{i+1} - x_i| = \Delta_n = 1/n$ and macroscale 1, which is length of the interval $(0, 1)$, we introduce a mesoscale by

$$\Delta'_n = R\Delta_n. \tag{11.39}$$

We partition the set of all points $\{x_i\}$ into n/R mesoscopic groups $G_q = \{x_{qR}, x_{qR+1} \ldots x_{(q+1)R}\}$, where $q = 0, 1, \ldots, n/R - 1$, and each group contains $R + 1$ points (■ Figure 11.7).

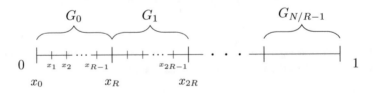

■ **Fig. 11.7** Partition into mesoscopic groups of R points

Step 2 (Decomposition of the Total Energy $E_n(\{u_i\})$)
The discrete energy of interactions within the qth group is

$$\mathcal{E}_n^{(q)} = \sum_{l=1}^{K} \sum_{i=qR}^{(q+1)R-l} \Delta_n \phi_l \left(\frac{u_{i+l} - u_i}{l\Delta_n} \right). \tag{11.40}$$

Besides interacting within their own group, points of neighboring groups may interact between each other. For example, the rightmost point of the $(q-1)$th group interacts the leftmost point of the qth group. The discrete energy of interaction between the $(q-1)$th and the qth group for each $q = 1, \ldots, n/R - 1$ is

$$\mathcal{E}_n^{(q-1,q)} = \sum_{l=1}^{K} \sum_{i=qR-l}^{qR} \Delta_n \phi_l \left(\frac{u_{i+l} - u_i}{l\Delta_n} \right). \tag{11.41}$$

Thus, the total discrete energy E_n given by (11.2) can be written as a sum of energies within each group and interactions between neighboring groups which results in a decomposition of the energy E_n into the sum

$$E_n(\{u_i\}) = \sum_{q=0}^{n/R-1} \mathscr{E}_n^{(q)} + \sum_{q=1}^{n/R-1} \mathscr{E}_n^{(q-1,q)} =: E_n^{(1)} + E_n^{(2)}. \tag{11.42}$$

The term $E_n^{(1)}$ represents the sum of energies due to the interactions within groups, while the term $E_n^{(2)}$ corresponds to interactions between the neighboring groups, that is $E_n^{(2)}$ describes the coupling between neighboring groups. The heuristic idea behind the partition of $\{x_i\}$ into groups $\{G_q\}$ and the energy decomposition (11.42) is to single out the dominating energy $E_n^{(1)}$ which consists of decoupled terms $\mathscr{E}_n^{(q)}$ and show that $E_n^{(2)}$ provides negligibly small contribution to the total energy. Indeed, one can expect that $E_n^{(1)}$ dominates when R is sufficiently large because the number of terms in $E_n^{(1)}$ is of the order n whereas the number of terms in $E_n^{(2)}$ is of the order n/R.

Step 3 (Derivation of the Cell Problem for Effective Lagrangian)
In view of Step 2, it is natural to consider minimization of $\mathscr{E}_n^{(q)}$ inside each group $G_q = \{x_{qR}, x_{qR+1}, \ldots, x_{(q+1)R}\}$ subject to prescribed values u_{qR} and $u_{(q+1)R}$ at the endpoints of the group. We next apply a simple scaling argument to obtain the expanding cell problem (11.43). This argument basically amounts to a shift and rescaling by Δ_n so that each group is treated in the same way, that is we eliminate the index q that numbers groups, and thus each group is replaced by a representative group parametrized by the slope α which enters through the boundary conditions, see (11.43).

Set $v_i := (u_{qR+i} - u_{qR})/\Delta_n$, and introduce $\alpha = v_R/R$, which is the slope of the linear function that interpolates the values u_{qR} and $u_{(q+1)R}$ at the endpoints of the qth group. The minimization problem for $\dfrac{1}{\Delta_n'}\mathscr{E}_n^{(q)}$ for each $q = 1, ..n/R$ writes as follows:

$$\overline{\phi}_R(\alpha) := \frac{1}{R} \min \left\{ \sum_{l=1}^{K} \sum_{i=0}^{R-l} \phi_l \left(\frac{v_{i+l} - v_i}{l} \right) \,\Big|\, \{v_0, v_1 \ldots v_{R-1}\}, \; v_0 = 0, \; v_R = \alpha R \right\}. \tag{11.43}$$

Once we pass to the limit $R \to \infty$ in (11.43) we derive the homogenized Lagrangian $\overline{\phi}(\alpha) := \lim_{R\to\infty} \overline{\phi}_R(\alpha)$ and thus establish a continuum Γ-limit (yet to be rigorously justified).

Before we rigorously prove the Γ-limit, we establish existence of the limit as $R \to \infty$ of $\overline{\phi}_R(\alpha)$ defined by (11.43) and its properties. To this end, we need several technical results, among which Proposition 11.5 is the crucial one.

Proposition 11.2 *The following a priori bound on $\overline{\phi}_R(\alpha)$ holds*

$$\overline{\phi}_R(\alpha) \le K C_2(|\alpha|^p + 1). \tag{11.44}$$

Proof

A priori bound (11.44) is obtained by taking $v_i = i\alpha$ in (11.43) and using the growth condition (11.9).   □

Next two propositions provide estimates on displacements near the right and left ends of the group (tuple). This will allow us to show that interactions between groups are negligible which will be rigorously shown by estimates on energies near interfaces between the groups.

Proposition 11.3 *Let tuple $\{v_0 = 0, v_1, \ldots v_R = \alpha R\}$ be a minimizer of the cell problem (11.43). Then for every $\varepsilon > 0$ there exists $s_0 = s_0(\varepsilon) \leq C/\varepsilon$ (with C independent of R and ε) and $s \in \{R - s_0, \ldots R - K\}$ such that*

$$\sum_{l=1}^{K-1} |v_{s+l} - v_s|^p \leq \varepsilon R, \tag{11.45}$$

and the following inequalities hold:

$$|v_s - v_R| = |v_s - \alpha R| \leq C R^{1/p}/\varepsilon^{\frac{p-1}{p}} \quad and \quad |v_{s+K} - v_R| = |v_{s+K} - \alpha R| \leq C R^{1/p}/\varepsilon^{\frac{p-1}{p}}. \tag{11.46}$$

Proof

Indeed, if (11.45) is violated then, by (11.9),

$$\frac{1}{R} \sum_{i=s}^{s+K-1} \phi_1 (v_{i+1} - v_i) \geq A\varepsilon \text{ with } A > 0,$$

and if (11.45) is violated for each $s = R - s_0, R - s_0 + 1, \ldots, R - K$, we have

$$\frac{1}{R} \sum_{l=1}^{K} \sum_{i=R-s_0}^{R-l} \phi_l \left(\frac{v_{i+l} - v_i}{l} \right) \geq \frac{1}{R} \sum_{i=R-s_0}^{R-1} \phi_1 (v_{i+1} - v_i) \geq A s_0 \varepsilon / K,$$

This yields the bound $s_0 \leq C/\varepsilon$, thanks to (11.44).

To prove the first inequality in (11.46) we use Holder's inequality, the fact that $\phi_1(v_m - v_{m-1}) \geq C_1 |v_m - v_{m-1}|^p$ and (11.44):

$$|v_s - v_R| \leq \sum_{m=s}^{R-1} |v_{m+1} - v_m| \leq \left(\sum_{m=s}^{R-1} |v_{m+1} - v_m|^p \right)^{1/p} (R - s)^{\frac{p-1}{p}} \leq C R^{1/p}/\varepsilon^{\frac{p-1}{p}}, \tag{11.47}$$

The second inequality in (11.46) is proved analogously.   □

Next proposition is straightforward analog of Proposition 11.3.

Proposition 11.4 *Let tuple $\{v_0 = 0, v_1, \ldots v_R = \alpha R\}$ be a minimizer of the cell problem* (11.43). *Then for every ε there exists $r_0 = r_0(\varepsilon) \leq C/\varepsilon$ (with C independent of R and ε) and $r \in \{0, \ldots r_0\}$ such that*

$$\sum_{l=1}^{K-1} |v_{r+l} - v_r|^p \leq \varepsilon R. \tag{11.48}$$

and

$$|v_r - v_0| = |v_r| \leq C R^{1/p} / \varepsilon^{\frac{p-1}{p}}. \tag{11.49}$$

Next proposition is the most technically involved but it provides the key estimate on $\overline{\phi}_R(\alpha)$. Specifically, estimate (11.50) is the main ingredient in the proof of existence of $\lim_{R \to \infty} \overline{\phi}_R(\alpha)$ in Theorem 11.3 (see (11.57)).

Proposition 11.5 *Let $\varepsilon > 0$ be sufficiently small, and consider $\overline{\phi}_R$ defined by* (11.43). *Then for $M > R^2 > \varepsilon^{-4}$ we can construct a tuple $\{\hat{v}_0 = 0, \hat{v}_1, \ldots, \hat{v}_M = \alpha M\}$ that satisfies*

$$\frac{1}{M} \sum_{l=1}^{K} \sum_{i=0}^{M-l} \phi_l \left(\frac{\hat{v}_{i+l} - \hat{v}_i}{l} \right) \leq \overline{\phi}_R(\alpha) + C\varepsilon + \frac{C}{(R\varepsilon)^{(p-1)/p}},$$

and thus

$$\overline{\phi}_M(\alpha) \leq \overline{\phi}_R(\alpha) + C\varepsilon + \frac{C}{(R\varepsilon)^{(p-1)/p}}. \tag{11.50}$$

Moreover, inequality (11.50) *holds uniformly when α varies over any bounded set in $\mathbb{R}$.*

Proof

Step I: Construction of the Tuple $\{\hat{v}_i\}$. The heuristic idea behind the proof is to obtain the desired test tuple $\{\hat{v}_0 = 0, \hat{v}_1, \ldots \hat{v}_M = \alpha M\}$ to satisfy (11.50) by "gluing together" shifted copies of the tuple $\{v_0 = 0, v_1, \ldots v_R = \alpha R\}$ (which is a minimizer of the cell problem (11.43)). These copies are identical up to shifts (by integer multiples of $v_s - v_r$, see (11.51)). The main difficulty lies in estimating energy near interfaces between the copies since gluing adds new interactions (near interfaces) that were not present in original minimization problem (11.43). That is why we cut r elements from the left and $s - K$ from the right ends of these tuples, where $r \ll R$ and $(s - K) \ll R$. The test tuple is constructed from the large "cut" pieces and the energy of "cut-off" pieces is estimated to be small. Then one needs to estimate energy of r and $s - K$ elements near the interfaces (estimates (11.45) and (11.48) below). Propositions 11.3 and 11.4 provide bounds s_0 and r_0 for number of elements in cut-off pieces.

Let s and r be as in Propositions 11.3 and 11.4, we construct the first $N(q)$ elements of the tuple $\{\hat{v}_i\}$ by setting

$$\{\hat{v}_0, \ldots \hat{v}_{s+K}\} := \{v_0, \ldots v_{s+K}\},$$

$$\{\hat{v}_{s+K+1}, \ldots \hat{v}_{2(s+K)+1-r}\} := \{v_r, \ldots v_{s+K}\} + (v_s - v_r),$$

$$\{\hat{v}_{2(s+K)+2-r}, \ldots \hat{v}_{3(s+K)+2-2r}\} := \{v_r, \ldots v_{s+K}\} + 2(v_s - v_r), \qquad (11.51)$$

$$\ldots$$

$$\{\hat{v}_{q(s+K)+q-(q-1)r}, \ldots \hat{v}_{(q+1)(s+K)+q-qr}\} := \{v_r, \ldots v_{s+K}\} + q(v_s - v_r),$$

where $N(q) := (q+1)(s+K+1) - qr$ (we have $|N(q) - (q+1)R| < Cq/\varepsilon$ with C independent of ε, R, and M), and q is an auxiliary positive integer parameter, to be chosen later. Also in the second line of (11.51) (and similarly in other lines), we use the notation $\{v_r, \ldots v_{s+K}\} + (v_s - v_r) = \{v_s, v_{s+1} + v_s - v_r, \ldots v_{s+K} + v_s - v_r\}$ for the shift by $(v_s - v_r)$. The energy of each tuple on the right-hand side of (11.51) is controlled by (11.43). The variation of the last K elements in the first line in (11.51) is controlled due to estimate (11.45), which is need to bound interaction energy between elements of the first and the second line when "glue" them together. This process repeats iteratively.

For $i \geq N(q)$ the values v_i are defined as a linear discrete function that matches the endpoint values $\hat{v}_{N(q)}$ and $\hat{v}_M = \alpha M$, i.e., $\hat{v}_i = v_{s+K} + qv_s - qv_r + \theta(i - N(q) + 1)$ for $i = N(q), \ldots M$ with

$$\theta := \frac{\hat{v}_M - \hat{v}_{N(q)}}{M - N(q)} = \frac{\alpha M - \hat{v}_{N(q)}}{M - N(q)}$$

and

$$q := \left[\frac{M}{R} - \frac{2M}{R^{2-1/p} \varepsilon^{(p-1)/p}} \right], \qquad (11.52)$$

where $[\cdot]$ denotes the integer part. With this choice of q the slope θ is bounded (uniformly) for $M > R^2 > \varepsilon^{-4}$. Indeed, since $|N(q) - (q+1)R| < Cq/\varepsilon$, it follows from (11.46) and (11.49) that

$$|\hat{v}_M - \hat{v}_{N(q)}| \leq Cq R^{1/p}/\varepsilon^{(p-1)/p}$$

and by (11.52) we have

$$M - N(q) \geq q R^{1/p}/\varepsilon^{(p-1)/p}. \qquad (11.53)$$

Step 2: Derivation of (11.50). The total energy (multiplied by M) of the tuple $\{\hat{v}_0 = 0, \hat{v}_1, \ldots \hat{v}_M = \alpha M\}$ can be represented as follows:

$$\sum_{l=1}^{K} \sum_{i=0}^{M-l} \phi_l \left(\frac{\hat{v}_{i+l} - \hat{v}_i}{l} \right) = \sum_{j=1}^{q+1} \left(\hat{E}_j^{\mathrm{I}} + \hat{E}_j^{\mathrm{II}} \right) + \hat{E}^{(\mathrm{lin})}, \tag{11.54}$$

where $\hat{E}_j^{\mathrm{I}} = \sum_{l=1}^{K} \sum_{i=t_j}^{T_j-K-1} \phi_l \left(\frac{\hat{v}_{i+l} - \hat{v}_i}{l} \right)$, $\hat{E}_j^{\mathrm{II}} = \sum_{l=1}^{K} \sum_{i=T_j-K}^{T_j} \phi_l \left(\frac{\hat{v}_{i+l} - \hat{v}_i}{l} \right)$, $\hat{E}^{(\mathrm{lin})} = \sum_{l=1}^{K} \sum_{i=N(q)-1}^{M-l} \phi_l \left(\frac{\hat{v}_{i+l} - \hat{v}_i}{l} \right)$. Here t_j and T_j are sub-indexes of the first and the last elements of the left-hand side tuple in the jth line of (11.51), i.e.,

$$
\begin{aligned}
t_1 &= 0, & T_1 &= s + K, \\
t_2 &= s + K + 1, & T_2 &= 2(s + K) + 1 - r, \\
t_3 &= 2(s + K) + 2 - r, & T_3 &= 3(s + K) + 2 - 2r, \\
&\cdots \\
t_{q+1} &= q(s + K) + q - (q - 1)r, & T_{q+1} &= (q + 1)(s + K) + q - qr.
\end{aligned}
$$

Due to the construction (11.51) of the tuple $\{\hat{v}_i\}$, the following bound holds:

$$\sum_{l=1}^{K} \sum_{i=0}^{M-l} \phi_l \left(\frac{\hat{v}_{i+l} - \hat{v}_i}{l} \right) \leq (q + 1)R(\overline{\phi}_R(\alpha) + C\varepsilon) + K(M - N(q))C_2(|\theta|^p + 1)$$

$$\leq R(q + 1)(\overline{\phi}_R(\alpha) + C\varepsilon) + \frac{C_3 M}{R^{1-1/p} \, \varepsilon^{(p-1)/p}}, \tag{11.55}$$

where we used the face that $\{v_0, \ldots, v_R\}$ is a minimizer of the cell problem (11.43) to bound $\hat{E}_j^{\mathrm{I}}$, estimates (11.48) and (11.45) to bound $\hat{E}_j^{\mathrm{II}}$, and (11.9) to bound $\hat{E}^{(\mathrm{lin})}$, i.e.,

$$\hat{E}_j^{\mathrm{I}} \leq R\overline{\phi}_R(\alpha), \quad \hat{E}_j^{\mathrm{II}} \leq CR\varepsilon, \quad \hat{E}^{(\mathrm{lin})} \leq K(M - N(q))C_2(|\theta|^p + 1).$$

For the second inequality in (11.55) we used (11.53).

Dividing (11.55) by M we obtain the required bound (11.50). Moreover, it is easily seen from the above proof that inequality (11.50) holds uniformly when α varies over any bounded set in $\mathbb{R}$. □

In the next theorem we prove existence of the homogenized Lagrangian and establish its properties.

Theorem 11.3 (homogenized Lagrangian is the limit energy over the expanding cell)

Let $\phi_l, l = 1, \ldots, K$ be continuous functions satisfying $C_1|\alpha|^p \leq \phi_l(\alpha) \leq C_2(|\alpha|^p + 1) \; \forall \alpha \in \mathbb{R}$ for some $0 < C_1 < C_2$. Then the following limit exists:

$$\overline{\phi}(\alpha) := \lim_{R \to \infty} \overline{\phi}_R(\alpha) \tag{11.56}$$

(continued)

Theorem 11.3 (continued)

$\forall \alpha \in \mathbb{R}$, and

(i) the inequality $\overline{\phi}_R(\alpha) \geq \overline{\phi}(\alpha) - \gamma(R, \alpha)$ holds with $\gamma(R, \alpha) \to 0$ as $R \to \infty$ uniformly with respect to α on bounded sets in $\mathbb{R}$;

(ii) $C_1|\alpha|^p \leq \overline{\phi}(\alpha) \leq K C_2(|\alpha|^p + 1)$;

(iii) $\overline{\phi}$ is convex.

Proof

Taking $\limsup_{M \to \infty}$ in (11.50), then $\liminf_{R \to \infty}$, and finally taking $\lim_{\varepsilon \to 0}$, we arrive at

$$\limsup_{M \to \infty} \overline{\phi}_M(\alpha) \leq \liminf_{R \to \infty} \overline{\phi}_R(\alpha), \tag{11.57}$$

that is $\exists \lim_{R \to \infty} \overline{\phi}_R(\alpha) =: \overline{\phi}(\alpha)$. This completes the proof of existence of the limit in (11.56).

Since according to Proposition 11.5 inequality (11.50) holds uniformly when α varies over any bounded set in $\mathbb{R}$, we obtain item (i).

The upper bound in item (ii) follows from inequality (11.44), while the lower bound is obtained as follows:

$$\overline{\phi}_R(\alpha) \geq \frac{1}{R} \sum_{i=0}^{R-1} \phi_1(v_{i+1} - v_i) \geq \frac{C_1}{R} \sum_{i=0}^{R-1} |v_{i+1} - v_i|^p$$

$$\geq C_1 \left(\frac{1}{R} \sum_{i=0}^{R-1} |v_{i+1} - v_i| \right)^p \geq C_1 |\alpha|^p, \tag{11.58}$$

where we used Holder's inequality.

Now we provide the proof of item (iii) (convexity). For every $M \geq R$ consider an auxiliary minimization problem

$$\overline{\Phi}_M(\alpha) := \frac{1}{M} \min \left\{ \sum_{l=1}^{K} \sum_{i=0}^{M-l} \phi_l \left(\frac{v_{i+l} - v_i}{l} \right) \, \Big| \, v_0 = \ldots v_K = 0, \right.$$

$$\left. v_{M-K} = \cdots = v_M = \alpha M \right\}. \tag{11.59}$$

It can be viewed as an analog of the cell problem (11.43) with additional boundary conditions. It is obvious that $\overline{\Phi}_M(\alpha) \geq \overline{\phi}_M(\alpha)$, since (11.43) has less constraints than (11.59). We will show that

$$\lim_{M \to \infty} \overline{\Phi}_M(\alpha) = \lim_{R \to \infty} \overline{\phi}_R(\alpha). \tag{11.60}$$

Fix an arbitrary $\varepsilon > 0$ and set

$$R := M - [M^{\frac{1}{p}} / \varepsilon^{\frac{p-1}{p}}].$$

Consider a minimizing tuple $\{v_0 = 0, v_1, \ldots v_R = \alpha R\}$ for the problem (11.43). In order to obtain an appropriate bound for (11.59), we construct a test tuple $\{\hat{v}_1, \ldots \hat{v}_M\}$. Take $r \in \{0, \ldots s_0(\varepsilon)\}$ and $s \in \{R - s_0(\varepsilon), \ldots R - K\}$ from Propositions 11.3 and 11.4, and set

$$\hat{v}_0 = \cdots = \hat{v}_K = 0,$$

$$\hat{v}_{K+1} = v_{r+1} - v_r, \ldots \hat{v}_{s+2K-r} = v_{s+K} - v_r.$$

The remaining elements in the tuple $\{\hat{v}_1, \ldots \hat{v}_M\}$ are constructed linearly

$$\hat{v}_{s+2K-r+i} = v_{s+K} + i\theta \text{ for } i = 1, \ldots (M - 3K + r - s - 1),$$

except the last $K + 1$ elements that are set equal to αM (to satisfy constraints in (11.59)):

$$\hat{v}_{M-K} = \cdots = \hat{v}_M = \alpha M.$$

Repeating the same arguments as in the proof of Proposition 11.5 we obtain from (11.46) and (11.49) that the slope θ is uniformly (in M) bounded, and using (11.45) and (11.48) we derive an analog of the bound (11.50)

$$\overline{\Phi}_M(\alpha) \leq \overline{\phi}_R(\alpha) + C\varepsilon + \frac{C}{(R\varepsilon)^{(p-1)/p}}. \tag{11.61}$$

The equality (11.60) is then obtained by taking $\limsup_{M\to\infty}$ and $\lim_{\varepsilon\to 0}$ of (11.61). Moreover, similarly to $\overline{\phi}_R(\alpha)$, the convergence of $\overline{\Phi}(\alpha)$ is uniform with respect to α on bounded sets in $\mathbb{R}$.

The established equality (11.60) means that $\overline{\phi}(\alpha)$ can be calculated as the limit of $\Phi_M(\alpha)$ as $M \to \infty$. We use this fact to prove convexity of $\overline{\phi}(\alpha)$. Namely, consider for given α, β and $\lambda \in (0, 1)$ a minimizer $\{\hat{v}_i\}$ for $\overline{\Phi}_{[\lambda R]}\left(\frac{\lambda\alpha M}{[\lambda R]}\right)$ and a minimizer $\{\tilde{v}_i\}$ for $\overline{\Phi}_{[(1-\lambda)R]}\left(\frac{(1-\lambda)\beta M}{[(1-\lambda)R]}\right)$, where $M = M(R) := [\lambda R] + [(1 - \lambda)R]$. Constructing the tuple $\{\hat{v}_0, \ldots \hat{v}_{[\lambda R]}, \tilde{v}_1 + \hat{v}_{[\lambda R]}, \ldots \tilde{v}_{[(1-\lambda)R]} + \hat{v}_{[\lambda R]}\}$ we get

$$\overline{\Phi}_{M(R)}(\lambda\alpha + (1 - \lambda)\beta) \leq \frac{[\lambda R]}{M(R)} \overline{\Phi}_{[\lambda R]}\left(\frac{\lambda\alpha M(R)}{[\lambda R]}\right)$$

$$+ \frac{[(1 - \lambda)R]}{M(R)} \overline{\Phi}_{[(1-\lambda)R]}\left(\frac{(1 - \lambda)\beta M(R)}{[(1 - \lambda)R]}\right).$$

Passing to the limit in this relation as $R \to \infty$ and using the fact that the convergence of $\overline{\Phi}(\alpha)$ is uniform with respect to α on bounded sets in $\mathbb{R}$ we get $\overline{\phi}(\lambda\alpha + (1 - \lambda)\beta) \leq \lambda\overline{\phi}(\alpha) + (1 - \lambda)\overline{\phi}(\beta)$ and the theorem is proved. $\square$

❶ Remark 11.6 Using minimizers of (11.59) one constructs piecewise linear test functions $w_{n,R,\alpha}(x)$ on interval of length $\Delta'_n := R\Delta_n$ by liner interpolating values $v_0\Delta_n, ..., v_R\Delta_n$ at points $x_0, ..., x_R$, where $\{v_0, ..., v_R\}$ is a minimizing tuple of problem (11.59). Then

$$\sum_{l=1}^{K}\int_0^{\Delta'_n - \Delta_n(l-1)} \phi_l\left(\frac{1}{l}\sum_{m=0}^{l-1} w'_{n,R,\alpha}(x + m\Delta_n)\right)dx = \Delta'_n \overline{\Phi}_R(\alpha) = \Delta'_n(\overline{\phi}_R(\alpha) + o(1)),$$

$$(11.62)$$

and

$$w_{n,R,\alpha} = 0 \text{ on } [0, (K-1)\Delta_n] \quad \text{and } w_{n,R,\alpha} = \alpha\Delta'_n \text{ on } [\Delta'_n - (K-1)\Delta_n, \Delta'_n]. \quad (11.63)$$

These functions will be used in the construction of the recovery sequence in the following theorem.

Theorem 11.4 (Homogenization of Nonconvex Problems with Long Range Interactions)

Let (11.9) hold with continuous functions ϕ_l. Then the functionals $E_n^{(c)}$ (given by (11.7)) Γ-converge in $L^p(0, 1)$ to the functional

$$E(u) = \begin{cases} \int_0^1 \overline{\phi}(u')dx & \text{if } u \in W^{1,p}(0, 1) \\ +\infty, & \text{otherwise,} \end{cases} \qquad (11.64)$$

where $\overline{\phi}$ is given by the following limit of minimization problems over expanding cells:

$$\overline{\phi}(\alpha) := \lim_{R\to\infty} \frac{1}{R}\min\left\{\sum_{l=1}^{K}\sum_{i=0}^{R-l} \phi_l\left(\frac{v_{i+l} - v_i}{l}\right)\,\Big|\,\{v_0, v_1 \ldots v_{R-1}\},\ v_0 = 0,\ v_R = \alpha R\right\}.$$

$$(11.65)$$

Existence of this limit is guaranteed by Theorem 11.3.

Proof

As in Theorem 11.1 it suffices to check the lim inf inequality for every sequence of piecewise linear functions $u_n \in \mathcal{L}_n$ such that $u_n \rightharpoonup u$ weakly in $W^{1,p}(0, 1)$. For every integer $R \geq 1$ we have,

$$E_n^{(c)}(u_n) \geq \sum_{q=0}^{[n/R]-1} \sum_{l=1}^{K} \sum_{i=qR}^{(q+1)R-l} \Delta_n \phi_l \left(\frac{u(x_{i+l})/\Delta_n - u(x_i)/\Delta_n}{l} \right)$$

$$\geq \sum_{q=0}^{[n/R]-1} R \Delta_n \overline{\phi}_R \left(\frac{u(x_{(q+1)R}) - u(x_{qR})}{R\Delta_n} \right) = \int_0^{[n/R]R\Delta_n} \overline{\phi}_R(\tilde{u}'_{R,n}(x)) dx$$

$$\geq \int_0^{[n/R]R\Delta_n} \overline{\phi}_R^{**}(\tilde{u}'_{R,n}(x)) dx,$$

where $\tilde{u}_{R,n}$ is the piecewise linear interpolation of values $u(x_0), u(x_R), \ldots u(x_{[n/R]R})$. Since $\tilde{u}_{R,n} \rightharpoonup u$ weakly in $W^{1,p}(0, b)$ on every interval $[0, b]$ $\forall b < 1$ (R is fixed), we obtain the following bound:

$$E_n^{(c)}(u_n) \geq \int_0^1 \overline{\phi}_R^{**}(u'(x)) dx. \tag{11.66}$$

By convexity of $\overline{\phi}$, Theorem 11.3 (i) and (11.58) the inequality $\liminf_{R\to\infty} \overline{\phi}_R^{**}(\alpha) \geq \overline{\phi}(\alpha)$ holds $\forall \alpha \in \mathbb{R}$. Therefore taking $\liminf_{R\to\infty}$ in the right-hand side of (11.66) we conclude that the first (lim inf) condition for Γ-convergence is satisfied, by applying Fatou's lemma.

The lim sup inequality is proved for piecewise linear functions $u(x)$. This is sufficient for establishing the lim sup inequality in general case by the continuity of $E(u)$ on $W^{1,p}(0, 1)$ and density of piecewise linear functions in $W^{1,p}(0, 1)$. Choose an integer $R \geq 1$. We construct $u_{R,n}(x)$ by setting $u_{R,n}(x) := u(x_{Rq}) + w_{n,R,\alpha_q}(x)$ on each interval $[q\Delta'_n, (q+1)\Delta'_n] \subset [0, 1]$ where $\alpha_q := u'(x)$ is constant (here $\Delta'_n = R\Delta_n$ and $q \in \mathbb{Z}$). Otherwise, if $u'(x)$ has a jump in the interval $[q\Delta'_n, (q+1)\Delta'_n]$, then we construct $u_{R,n}(x)$ by linearly interpolating values of u on the ends of the interval.

Next we show that

$$u_{R,n} \text{ weakly converges to } u \text{ in } W^{1,p}(0, 1) \text{ as } n \to \infty \tag{11.67}$$

and

$$\limsup_{n\to\infty} E_n^{(c)}(u_{R,n}) \leq \int_0^1 \overline{\Phi}_R(u'(x)) dx + C/R. \tag{11.68}$$

To prove (11.67) use growth condition (11.9)

$$C_1 \int_0^1 |u'_{R,n}(x)|^p \, dx \leq \sum_{q=0}^{[n/R]-1} \int_{q\Delta'_n}^{(q+1)\Delta'_n} \phi_1\left(u'_{R,n}(x)\right) \, dx$$

$$= \sum_{q=0}^{[n/R]-1}{}' \int_{q\Delta'_n}^{(q+1)\Delta'_n} \phi_1\left(w'_{R,n,u'(x)}(x)\right) \, dx + o(1), \tag{11.69}$$

where $\sum'$ means summation over those q such that $u'(x)$ does not have a jump in the interval $[q \Delta_n', (q+1)\Delta_n']$, the remaining terms (which correspond to intervals where $u'(x)$ has a jump; number of such intervals is vanishingly small as $n \to \infty$) contribute to the $o(1)$ term. Using (11.62) we obtain that $u_{R,n}$ is bounded in $W^{1,p}(0,1)$ and since $u_{R,n}$ converges to u pointwise ($u_{R,n}(q\Delta_n') = u(q\Delta_n')$) we obtain (11.67).

To prove (11.68) represent the total energy as follows:

$$E_n^{(c)}(u_{R,n}) = E_n(\{u_{R,n}(x_i)\}) = \sum_{q=0}^{[n/R]-1}{}' \mathscr{E}^{(q)} + \sum_{q}{}'' \mathscr{E}^{(q)} + \sum_{q=1}^{[n/R]-1} \mathscr{E}^{(q-1,q)} + \mathscr{E}^{\text{rem}},$$

(11.70)

where $\mathscr{E}^{(q)}$ and $\mathscr{E}^{(q-1,q)}$ are defined in (11.40) and (11.41). Here $\sum''$ means summation over those q such that $u'(x)$ has a jump in the interval $[q\Delta_n', (q+1)\Delta_n']$. The term $\mathscr{E}^{\text{rem}}$ is due to n/R may not be an integer, so it takes into account the energy of elements outside any group of R elements:

$$\mathscr{E}^{\text{rem}} = \sum_{l=1}^{K} \sum_{i=[n/R]R}^{n-l} \Delta_n \phi_l \left(\frac{u(x_{i+l}) - u(x_i)}{l\Delta_n} \right).$$

Each term $\mathscr{E}^{(q)}$ is bounded by $\Delta_n' \overline{\Phi}_R(\alpha_q)$, whereas three other terms in the right-hand side of (11.70) are bounded by C/R. This proves (11.68).

Finally it remains to choose appropriately $R(n)$ ($R(n) \to \infty$ as $n \to \infty$) to obtain a sequence $u_n = u_{R(n),n}$ with the required properties (weakly converging to u in $W^{1,p}(0,1)$ and such that $\limsup_{n\to\infty} E_n^{(c)}(u_n) \leq E(u)$), i.e., $\{u_n\}$ is indeed a recovery sequence. Theorem is proved. □

11.4 Expanding Cell Problems vs Periodic Cell Problem

In this subsection we show that in the special case $K = 2$, the expanding cell problem for tuples of size $R \gg K$ is equivalent to periodic cell problem so the homogenized Lagrangian can be defined via the periodic cell problem. Then the homogenization is analogous to $K = 1$ in the following sense. The homogenized Lagrangian is obtained by taking convex envelope of not ϕ but of the minimal energy of a periodic cell problem (see (11.71) below). We also show by counterexample that this simplification does not hold for $K \geq 3$.

Note that the discrete energy functional (11.2) is K periodic (away from the boundary), which is a clear analogy with the continuum periodic problems considered in ▶ Section 9. Using this analogy one can write the following discrete $K-$periodic cell problem:

$$\phi_{per}(\alpha) = \frac{1}{K} \min \left\{ \sum_{l=1}^{K} \sum_{i=0}^{K-1} \phi_l \left(\frac{v_{i+l} - v_i}{l} \right) : v_0 = 0, \ v_{i+K} = v_i + K\alpha \right\}.$$

(11.71)

Proposition 11.6 *Let $\bar{\phi}$ be homogenized Lagrangian given by Lemma 11.3. Then*

$$\bar{\phi}(\alpha) \le \phi_{per}(\alpha) \le \sum_{l=1}^{K} \phi_l(\alpha). \tag{11.72}$$

Moreover, if ϕ_l are convex, then

$$\bar{\phi}(\alpha) = \phi_{per}(\alpha). \tag{11.73}$$

Proof

Taking $v_i = \alpha i$ in the right-hand side if (11.71) we get that $\phi_{per}(\alpha) \le \sum_{l=1}^{K} \phi_l(\alpha)$. On the other hand, every minimizer of (11.71) can be continued quasi-periodically $v_{i+K} = v_i + K\alpha$ to get a test tuple for problem (11.43), and this leads to the inequality $\phi_{per}(\alpha) \ge \bar{\phi}(\alpha)$. Thus, (11.72) is proven. Finally, if we assume that ϕ_l are convex, then by Theorem 11.1 we have that $\bar{\phi}(\alpha) = \sum_{l=1}^{K} \phi_l(\alpha)$ which implies (11.73). □

Proposition 11.6 states that inequalities (11.72) always hold, even for nonconvex energies. Moreover in particular cases of nearest neighbor and next to nearest neighbor interactions, $K = 1, 2$, the upper bound provided by $\phi_{per}(\alpha)$ is optimal in the sense that $\bar{\phi}(\alpha) = \phi_{per}^{**}(\alpha)$. For $K = 1$ this was proved in Theorem 11.2. This property is also valid for $K = 2$ as shown in the following lemma.

ⓘ Lemma 11.1 *Assume $K = 2$, then*

$$\bar{\phi}(\alpha) = \phi_{per}^{**}(\alpha). \tag{11.74}$$

Proof

Step 1. Let R be even. Consider a minimizer $\{v_i\}$ of (11.71) for $K = 2$, continued to the tuple $\{v_0, \ldots v_R\}$ by setting $v_{i+2} = v_i + 2\alpha$. Then we have $v_0 = 0$, $v_R = \alpha R$, hence by (11.43)

$$R\bar{\phi}_R(\alpha) \le \sum_{l=1}^{2} \sum_{i=0}^{R-l} \phi_l\left(\frac{v_{i+l} - v_i}{l}\right) = \sum_{l=1}^{2} \sum_{i=0}^{R-2} \phi_l\left(\frac{v_{i+l} - v_i}{l}\right) + \phi_1(v_R - v_{R-1})$$

$$= \sum_{q=0}^{R/2-2} \sum_{i=2q}^{2q+1} \sum_{l=1}^{2} \phi_l\left(\frac{v_{i+l} - v_i}{l}\right) + \phi_1(v_{R-1} - v_{R-2}) + \phi_1(v_R - v_{R-1})$$

$$+ \phi_2\left(\frac{v_{R-2} - v_R}{2}\right) \le 2 \sum_{q=0}^{R/2-2} \phi_{per}(\alpha) + 2\phi_{per}(\alpha) = R\phi_{per}(\alpha).$$

Thus $\bar{\phi}(\alpha) = \lim_{R \to \infty} \bar{\phi}_R(\alpha) \le \phi_{per}(\alpha)$, and since $\bar{\phi}(\alpha)$ is convex, we have $\bar{\phi}(\alpha) \le \phi_{per}^{**}(\alpha)$.

Step 2. In order to show the reversed inequality $\overline{\phi}(\alpha) \geq \phi_{per}^{**}(\alpha)$ we remind that in the proof of Lemma 11.3 we have established the representation $\overline{\phi}(\alpha) = \lim_{M\to\infty} \overline{\Phi}_M(\alpha)$, where $\overline{\Phi}_M(\alpha)$ is given by (11.59). For even M consider a minimizer $\{v_0, \ldots v_M\}$ of (11.59). We have, after rearranging terms in sums,

$$
M\overline{\Phi}_M(\alpha) = \phi_1(0)
$$

$$
+ \sum_{q=0}^{M/2-1} \left(\phi_2\left(\frac{v_{2q+2} - v_{2q}}{2}\right) + \frac{1}{2}\left(\phi_1(v_{2q+1} - v_{2q}) + \phi_1(v_{2q+2} - v_{2q+1})\right)\right)
$$

$$
+ \sum_{q=0}^{M/2-2} \left(\phi_2\left(\frac{v_{2q+3} - v_{2q+1}}{2}\right) + \frac{1}{2}\left(\phi_1(v_{2q+2} - v_{2q+1}) + \phi_1(v_{2q+3} - v_{2q+2})\right)\right).
$$

$$(11.75)$$

Observe that each term of the form

$$
\phi_2\left(\frac{v_{i+2} - v_i}{2}\right) + \frac{1}{2}\left(\phi_1(v_{i+1} - v_i) + \phi_1(v_{i+2} - v_{i+1})\right)
$$

in (11.75) is bounded from below by $\phi_{per}((v_{i+2} - v_i)/2)$ (to see this, one can construct the test tuple $\{0, v_{i+1} - v_i, v_{i+2} - v_i, v_{i+2} + v_{i+1} - 2v_i\}$ in (11.71)). Thus,

$$
M\overline{\Phi}_M(\alpha) \geq \sum_{q=0}^{M/2-1} \phi_{per}\left(\frac{v_{2q+2} - v_{2q}}{2}\right) + \sum_{q=0}^{M/2-2} \phi_{per}\left(\frac{v_{2q+3} - v_{2q+1}}{2}\right) \geq
$$

$$
\geq \sum_{q=0}^{M/2-1} \phi_{per}^{**}\left(\frac{v_{2q+2} - v_{2q}}{2}\right) + \sum_{q=0}^{M/2-2} \phi_{per}^{**}\left(\frac{v_{2q+3} - v_{2q+1}}{2}\right).
$$

Then, due to the convexity of ϕ_{per}^{**} we conclude

$$
M\overline{\Phi}_M(\alpha) \geq \frac{M}{2}\phi_{per}^{**}\left(\frac{2}{M}\sum_{q=0}^{M/2-1} \frac{v_{2q+2} - v_{2q}}{2}\right)
$$

$$
+ \frac{M-2}{2}\phi_{per}^{**}\left(\frac{2}{M-2}\sum_{q=0}^{M/2-2} \frac{v_{2q+2} - v_{2q}}{2}\right)
$$

$$
= \frac{M}{2}\phi_{per}^{**}(\alpha) + \frac{M-2}{2}\phi_{per}^{**}(\frac{M}{M-2}\alpha).
$$

It remains to divide this inequality by M and pass to the limit as $M \to \infty$. Lemma is proved.
□

167

11

11.5 · Continuum Approximations of Discrete Minimization Problems via Γ-...

It is natural to ask if the simple way (11.74) of computing the homogenized Lagrangian holds for longer ranges of interactions that is for $K > 2$. Due to K-periodicity of the energy one can guess that minimizers of (11.43) have periodicity property $v_{i+K} = v_i + \alpha K$ or exhibit a close to K-periodic behavior locally on the microscale 1. This would lead to the formula $\overline{\phi}(\alpha) = \phi_{per}^{**}(\alpha)$ by convexification on the mesoscale R. The example 11.2 below shows that when $K > 2$, the proof of Lemma 11.1 no longer provides K-periodic "almost minimizers" that is the subtle interplay between the micro and mesoscales results in the failure of (11.74) for $K > 2$. Since the interaction range parameter K is intrinsic for discrete problems this example demonstrates an essential difference between continuum and discrete problems (cf [76]).

Example 11.2 (Failure of (11.74) for $K = 3$)
For $K = 3$ consider energies

$$\phi_1(s) = (s^2 - 1)^2(s^2 - 4)^2(s^2 - 16)^2 + 1,$$

$$\phi_2(s) = (s^2 - 9/4)^6 + 1, \tag{11.76}$$

$$\phi_3(s) = (s^2 - 1/9)^2 (s^2 - 4/9)^2 (s^2 - 16/9)^2 + 1.$$

It follows from (11.71) and (11.76) that $\phi_{per}(\alpha) \geq 3$ ($\forall \alpha \ \phi_i(\alpha) \geq 1, i = 1, 2, 3$) and if $\phi_{per}(\alpha) = 3$, then there exists a tuple $\{v_0 = 0, v_1, v_2, v_3 = 3\alpha, v_4, v_5\}$ such that $v_{i+3} = v_i + 3\alpha \ \forall i$, and $\phi_l(\frac{v_{i+l}-v_i}{l}) = 1, i = 0, 1, 2, l = 1, 2, 3$. This latter condition yields:
1. $v_{i+1} - v_i \in \{\pm1, \pm2, \pm4\}$;
2. $v_{i+2} - v_i = \pm3$;
3. $v_{i+3} - v_i \in \{\pm1, \pm2, \pm4\}$.

Exhausting all possible combinations satisfying conditions 1.-3. and the quasi-periodicity condition $v_{i+3} = v_i + 3\alpha \ \forall i$ shows that such a tuple does not exist. Thus $\phi_{per}(\alpha) > 3$ for all α, therefore $\min_\alpha \phi_{per}(\alpha) > 3$ by continuity of $\phi_{per}(\alpha)$ and growth conditions $\phi_l(s) \geq C_1|s|^{12}$. On the other hand one can construct a 4-periodic tuple $\{v_i\}$ ($v_{i+4} = v_i \ \forall i$) such that $\phi_l(\frac{v_{i+l}-v_i}{l}) = 1$ for all $l = 1, 2, 3$ and $i = 0, 1 \ldots$. One of such 4-periodic tuples is given by $v_0 = 0, v_1 = 1, v_2 = 3, v_3 = 4, v_4 = 0, \ldots$. Substitute this tuple in (11.43) for $\alpha = 0$ and R multiple 4, to get $\overline{\phi}(0) = \lim_{R\to\infty} \overline{\phi}_R(0) = 3$. Therefore we have $\overline{\phi}(0) < \min_\alpha \phi_{per}(\alpha) = \min_\alpha \phi^{**}(\alpha)$ and the formula (11.74) fails for $K = 3$. ∎

11.5 Continuum Approximations of Discrete Minimization Problems via Γ-Convergence

Finally, we return to minimization problem (11.8). The second term in (11.8) can be thought of as an integral sum of $\int_0^1 u_n(x) f(x) \, dx$. Moreover, one can show that under some a priori bounds on $\{f_i = f_i(n)\}$, e.g., $\frac{1}{n}\sum (f_i(n))^{p/(p-1)} \leq C$, we always can extract a subsequence such that

$$\sum_{i=0}^{n} f_i(n) u_i \Delta_n - \int_0^1 f(x) u_n(x) dx = o(\|u_n\|_{W^{1,p}(0,1)}) \text{ for some } f \in L^p(0,1),$$

(11.77)

where $u_n(x)$ is the piecewise linear interpolation of values u_i at points x_i. Then, assuming (11.77) and (11.9), we have

  if $E_n^{(c)}(u)$ Γ-converges to $E(u)$ in $L^p(0,1)$

  then $E_n^{(c)}(u) - \sum_{i=0}^{n} f_i(n) u_i \Delta_n$ Γ-converges to $E(u) - \int_0^1 f(x) u_n(x) dx$ in $L^p(0,1)$.

Also, revising proofs of Theorems 11.1, 11.2, 11.4 one sees that Γ-convergence results in these theorems are not affected by prescribed boundary conditions. That is, $L^p(0,1)$-convergence can be replaced by weak $W_0^{1,p}(0,1)$-convergence in Theorems 11.1, 11.2, 11.4. Then, by using general properties of Γ-convergence (Theorem 9.1) one arrives at

Theorem 11.5
Let (11.77) *hold together with conditions of Theorems 11.1 or Theorem 11.2 or Theorem 11.4. Then minima of problems* (11.8) *converge to the minimum of the problem*

$$\min_{u \in W^{1,p}(0,1)} \left\{ E(u) - \int_0^1 f(x) u(x) dx \right\},$$

(11.78)

moreover, minimizers of (11.8) *converge, up to a subsequence, to a minimizer of* (11.78).

Appendix A
Regular and Singular Perturbations and Boundary Layers

© Springer Nature Switzerland AG 2018
L. Berlyand, V. Rybalko, *Getting Acquainted with Homogenization and Multiscale*,
Compact Textbooks in Mathematics,
https://doi.org/10.1007/978-3-030-01777-4

In this appendix by means of two simple examples we illustrate what regular and singular perturbations mean.

A.1 Regular Perturbation

Consider the following initial value problem with a parameter ε:

$$y'_\varepsilon(t) = \varepsilon y_\varepsilon + b, \quad t \in [0, T], \tag{A.1}$$

$$y_\varepsilon(0) = c, \tag{A.2}$$

where $b \in \mathbb{R}$. The solution of this equation is

$$y_\varepsilon(t) = \left(c + \frac{b}{\varepsilon}\right) e^{\varepsilon t} - \frac{b}{\varepsilon} = c e^{\varepsilon t} - \frac{b}{\varepsilon}(1 - e^{\varepsilon t}). \tag{A.3}$$

The solutions $y_\varepsilon(t)$ converge uniformly to the limiting solution $y_0(t)$:

$$y_\varepsilon(t) = c e^{\varepsilon t} - \frac{b}{\varepsilon}(1 - e^{\varepsilon t}) \approx c e^{\varepsilon t} + \frac{b}{\varepsilon} \cdot \varepsilon t \rightrightarrows c + bt = y_0(t), \tag{A.4}$$

where $y_0(t)$ solves the problem (A.1)–(A.2) with $\varepsilon = 0$:

$$y'(t) = 0 \cdot y + b = b, \quad t \in [0, T], \tag{A.5}$$

$$y(0) = c, \tag{A.6}$$

In fact, (A.5)–(A.6) can be obtained as the "naive limit" by taking $\varepsilon \to 0$ in (A.1)–(A.2). The problem (A.1)–(A.2) is called a *regular perturbation* of the problem (A.5)–(A.6). A solution of (A.1)–(A.2) uniformly converges to the solution of (A.5)–(A.6) as $\varepsilon \to 0$ and (A.5)–(A.6) does not depend on any small parameter. Therefore to approximate the solution of (A.1)–(A.2) for small ε it is sufficient to solve the simpler problem obtained by substituting $\varepsilon = 0$.

A.2 Singular Perturbation

Now consider the following initial value problem where the parameter ε affects the highest derivative:

$$\varepsilon y'(t) = -ay + b, \quad t \in [0, T], \tag{A.7}$$

$$y(0) = c, \tag{A.8}$$

where $a > 0$, $b \in \mathbb{R}$. Notice that if $\varepsilon = 0$, (A.7) is no longer an ODE:

$$0 = -ay_0(t) + b \Rightarrow y_0(t) = \frac{b}{a} \text{ for any } t \in [0, T], \tag{A.9}$$

and, thus, the solution $y_0(t)$ does not depend on initial condition (A.8). At this point the difference between regular and singular perturbation is clearly visible. Regular perturbation solutions of the pre-limiting problem converge uniformly to the solution of the problem with $\varepsilon = 0$, while singular perturbation solutions lack uniform convergence if $c \neq b/a$.

Indeed, the solutions in the case $\varepsilon > 0$ can be found explicitly:

$$y_\varepsilon(t) = \left(c - \frac{b}{a}\right) e^{-\frac{a}{\varepsilon}t} + \frac{b}{a}. \tag{A.10}$$

These solutions start at c (when $t = 0$) and approach b/a when $t \to \infty$. Moreover,

$$\left(c - \frac{b}{a}\right) e^{-\frac{a}{\varepsilon}t} \to 0 \text{ when } \varepsilon \to 0 \tag{A.11}$$

for all $t > 0$, so at any fixed point $t > 0$ the solution $y_\varepsilon(t)$ converges to b/a (pointwise convergence). However $y_\varepsilon(0) = c$ for all ε and thus no uniform convergence holds.

The problem (A.7)–(A.8) is called a *singular perturbation* of the problem $0 = -ay + b$ with no small parameter ε, since the problem (A.7)–(A.8) has ε in front of the highest order term. The equation (A.7) in the case $\varepsilon = 0$ degenerates into a lower order differential equation that has a qualitatively different solution.

The first term in the right-hand side of (A.10) describes a so-called *boundary layer*. Specifically, divide the time interval $[0, T]$ into two regions: $[0, t_0]$ and $(t_0, T]$ such that the solution of the problem with $\varepsilon = 0$ is a good approximation for $y_\varepsilon(t)$ on $(t_0, T]$:

$$\sup_{t\in(t_0,T]} |y_\varepsilon(t) - y_0| \le \left(c - \frac{b}{a}\right) e^{-\frac{a}{\varepsilon}t_0} \to 0 \text{ as } \varepsilon \to 0.$$

In particular, for every $t_0 > 0$ we have $e^{-\frac{a}{\varepsilon}t_0} \to 0$ as $\varepsilon \to 0$. That is we have an exponential boundary layer at $t = 0$.

The width of the boundary layer $(0, t_0)$, outside which y_ε uniformly converges to a constant, can be taken, for example, as ε^α for any $\alpha \in (0, 1)$. Boundary layers originally appeared in fluid dynamics in the description of flow with small viscosity near walls (e.g., Prandtl equation).

References

1. Acerbi, E., Chiadò-Piat, V., Dal Maso, G., Percivale, D.: An extension theorem from connected sets, and homogenization in general periodic domains. Nonlinear Analysis: Theory, Methods and Applications **18**(5), 481–496 (1995)
2. Alberti, G., Müller, S.: A new approach to variational problems with multiple scales. Commun. Pure Appl. Math. **54**(7), 826–850 (2001)
3. Allaire, G.: Homogenization and two scale convergence. SIAM J. Math. Anal. **23** (6), 1482–1518 (1992)
4. Allaire, G.: Shape Optimization by the Homogenization Method. Springer (2010)
5. Arbogast, T., Douglas, J., Hornung, U.: Derivation of the double porosity model of single phase flow via homogenization theory. SIAM J. Math. Anal. **21**(4), 823–836(1990)
6. Armstrong, S., Caraliaguet P.: Stochastic homogenization of quasilinear Hamilton-Jacobi equations and geometric motions J. Eur. Math. Soc., to appear
7. Armstrong, S., Smart, C.: Quantitative stochastic homogenization of convex integral functionals. Ann. Sci. Éc. Norm. Supér. (4) **49**(2), 423–481 (2016)
8. Arnold, V. I.: Mathematical Methods of Classical Mechanics. Springer (1997)
9. Assyr, A., E, W., Engquist, B., Vanden-Eijnden E.: The heterogeneous multiscale method. Acta Numerica **21**, 1–87 (2012)
10. Babuška, I.: Solution of interface problems by homogenization. I,II. SIAM J. Math. Anal. **7**(5), 603–634, 635–645 (1976)
11. Babuška, I., Lipton, R.: Optimal local approximation spaces for generalized finite element methods with application to multiscale problems. Multiscale Model. Simul. **9**(1), 373–406 (2011)
12. Ball, J., James, R.: Fine phase mixtures as minimizers of energy. Arch. Rational Mech. Anal. **100**(1), 13–52 (1987)
13. Barabanov, O. O., Zhikov, V. V. The Limit load and homogenization. Izvestiya: Mathematics **43**(2), 205–231 (1994)
14. Barenblatt, G.I., Zheltov, I.P., Kochina, I.N.: Basic concepts in the theory of seepage of homogeneous liquids in fissured rocks, Journal of applied mathematics and mechanics **24**(5), 1286–1303 (1960)
15. Bakhvalov, N.S., Panasenko, G.: Homogenisation: Averaging Processes in Periodic Media. Springer (1989)
16. Bensoussan, A., Lions, J.-L., Papanicolaou, G.: Asymptotic analysis for periodic structures. North-Holland Pub. Co., Amsterdam (1978)
17. Braides, A., Defrancheschi, A.: Homogenization of Multiple Integrals (Oxford Lecture Series in Mathematics and Its Applications). Oxford University Press (1999)
18. Berlyand, L.V., Kozlov, S.M.: Asymptotics of the homogenized moduli for the elastic chess-board composite. Arch. Ration. Mech. Anal. **118** (2), 95–112 1992
19. Berlyand, L., Promislow, K. S.: Effective Elastic Moduli of a Soft Medium with Hard Polygonal Inclusions and Extremal Behavior of Effective Poisson's Ratio. Journal of Elasticity **40**(1), 45–73 (1995)
20. Berlyand, L., Kolpakov, A. G., Novikov, A.: Introduction to the Network Approximation Method for Materials Modeling. Cambridge University Press (2012)
21. Berlyand, L., Ohwadi, H.: Flux norm approach to finite dimensional homogenization approximations with non-separated scales and high contrast. Arch Rational Mech. Anal. **198**, 677–721 (2010)

© Springer Nature Switzerland AG 2018
L. Berlyand, V. Rybalko, *Getting Acquainted with Homogenization and Multiscale*,
Compact Textbooks in Mathematics,
https://doi.org/10.1007/978-3-030-01777-4

22. Berlyand, L. and Mityushev, V.: Increase and decrease of the effective conductivity of two phase composites due to polydispersity. J. Stat. Phys. **118**(3/4), 481–509 (2005)
23. Berlyand, L., Sandier, E., Serfaty, S.: A two scale Γ-convergence approach for random non-convex homogenization. Calc. Var. Partial Differ. Equ. **6**, 156 (2017)
24. Blanc, X., Le Bris, C., Legoll, F.: Some variance reduction methods for numerical stochastic homogenization. Philos. Trans. Roy. Soc. A **374** (2066), 20150168 (2016)
25. Bressan, A.: Lecture Notes on Functional Analysis: With Applications to Linear Partial Differential Equations. American Mathematical Society, vol. 143 (2013)
26. Bourgeat, A., Piatnitski, A.: Estimates in probability of the residual between the random and the homogenized solutions of one-dimensional second-order operator. Asymptot. Anal. **21**(3–4), 303–315 (1999)
27. Bourgeat, A., Piatnitski, A.: Approximations of effective coefficients in stochastic homogenization. Ann. Inst. H. Poincaré **40**(2), 153–165 (2004)
28. Braides, A.: Gamma-convergence for Beginners. Clarendon Press, Vol. 22 (2002)
29. Burago, D.: Periodic metrics. Advances in Soviet Mathematics **9**, 205–211 (1992)
30. Caffarelli, L.A., Souganidis, P.E.: Rates of convergence for the homogenization of fully nonlinear uniformly elliptic PDE in random media. Invent. Math. **180**(2), 301–360 (2010)
31. Caffarelli, L.A., Souganidis, P.E., Wang, L.: Homogenization of fully nonlinear, uniformly elliptic and parabolic partial differential equations in stationary ergodic media. Comm. Pure Appl. Math. **58**(3), 319–361 (2005)
32. Chakrabarty, J.: Theory of Plasticity. Elsevier Butterworth-Heinemann (2006)
33. Chechkin, G. A., Piatnitski, A. L., Shamev, A. S.: Homogenization: Methods and Applications. American Mathematical Soc., volume 234 (2007)
34. Cherkaev, A.: Variational Methods for Structural Optimization. Springer Science & Business Media, volume 140 (2012)
35. Chipot, M., Kinderlehrer, D.: Equilibrium configurations of crystals. Arch. Rational Mech. Anal. **103** (3), 237–277 (1988)
36. Christensen, R.M.: Mechanics of Composite Materials. Courier Corporation (2012)
37. Ciornescu, D., Donato, P.: An Introduction to Homogenization (Oxford Lecture Series in Mathematics and Its Applications). Oxford University Press (2000)
38. Cioranescu, D., Murat, F., A strange term coming from nowhere. Topics in the mathematical modelling of composite materials. In: Cherkaev A., Kohn R. (eds) Topics in the Mathematical Modelling of Composite Materials. Progress in Nonlinear Differential Equations and Their Applications, vol 31. Birkhäuser, Boston (1997)
39. Copson, E.T.: Asymptotic Expansions. Cambridge University Press, (No. 55) (2004)
40. Dal Maso, G.: An Introduction to Γ-convergence. Springer Science & Business Media, volume 8 (2012)
41. De Bruijn, N.G.: Asymptotic Methods in Analysis. Courier Corporation, vol 4. (1970)
42. Doob, J.L.: Stochastic Processes. John Wiley & Sons, Inc., New York; Chapman & Hall, Limited, London (1953)
43. Dychne, A.M.: Conductivity of a two-phase two-dimensional system. Journal of Experimental and Theoretical Physics **59**(7), 110–115 (1970).
44. E, W.: Principles of Multiscale Modeling. Cambridge University Press (2011)
45. Efendiev, Y., Hou, T.Y.: Multiscale Finite Element Methods: Theory and Applications. Springer Science & Business Media, volume 4 (2009)
46. Einstein, A.: Eine neue Bestimmung der Moleküldimensionen. Ann. Physik. **19**, 289–247 (1906)
47. Ekeland, I., Temem, R.: Convex Analysis and Variational Problems. Society for Industrial and Applied Mathematics (1987)
48. Engquist, B., Li, X., Ren, W., Vanden-Eijnden, E. and others: Heterogeneous multiscale methods: a review. Communications in Computational Physics **2**(3), 367–450 (2007)

49. Evans, L. C.: Partial Differential Equations: Second Edition (Graduate Studies in Mathematics). American Mathematical Society (2010)

50. Fish, J.: Multiscale Methods: Bridging the Scales in Science and Engineering. Oxford University Press on Demand (2010)

51. Freidlin, M. I.: Dirichlet's problem for an equation with periodical coefficients depending on a small parameter. Teor. Veroyatnost. i Primenen. **9**, 133–139 (1964)

52. Gilbarg, D., Trudinger, N. S.: Elliptic Partial Differential Equations of Second Order. Springer-Verlag Berlin Heidelberg (2001)

53. Gloria, A., Mourrat, J.-C.: Spectral measure and approximation of homogenized coefficients. Probab. Theory Related Fields **154** (1–2), 287–326 (2012)

54. Gloria, A., Otto, F.: An optimal variance estimate in stochastic homogenization of discrete elliptic equations. Ann. Probab. **39**(3), 779–856 (2011)

55. Gloria, A., Neukamm, S., Otto, F.: Quantification of ergodicity in stochastic homogenization: optimal bounds via spectral gap on Glauber dynamics. Invent. Math. **199**(2), 455–515 (2015)

56. Grimmett, G., Stirzaker, D.: Probability and Random Processes. Oxford university press (2001)

57. Gross, D., Hauger, W., Schröder, J., Wall, W., Bonet, J.: Engineering Mechanics 2: Mechanics of Materials. Springer (2011)

58. Hornung, U.: Homogenization and Porous Media. Springer (1997)

59. Hou, T. Y., Wu, X.-H.: A multiscale finite element method for elliptic problems in composite materials and porous media. Journal of computational physics **134**(1), 169–189 (1997)

60. Hou, T., Wu, X.-H., Cai, Z.: Convergence of a multiscale finite element method for elliptic problems with rapidly oscillating coefficients. Math. Comp. **68**(227), 913–943 (999)

61. Jeffery, D. J.: Conduction through a random suspension of spheres. Proc. R. Soc. Lond. A **335**, 335–367 (1973)

62. Jikov, V. V., Kozlov, S. M., Oleinik, O. A.: Homogenization of Differential Operators and Integral Functionals. Springer-Verlag (1994)

63. Kalamkarov, A. L., Kolpakov, A. G.: Analysis, Design, and Optimization of Composite Structures. Wiley (1997)

64. Keller, J. B.: A theorem on the conductivity of a composite medium. J. Math. Phys. **5**, 548–549 (1964)

65. Kolpakov, A. G.: Determination of the average characteristics of elastic frameworks. J. Appl. Math. Mech. **49**,(6), 739–745 (1985)

66. Kolpakov, A. G. Rakin, S. I.:Problem of the synthesis of a composite material of unidimensional structure with assigned characteristics. J. Appl. Mech. Tech. Phys. **27**(6), 917–922 (1986)

67. Levy, O., Kohn, R. V.: Duality relations for non-Ohmic composites, with applications to behavior near percolation. J. Stat. Phys. **90** (1), 159–189 (1998)

68. Målqvist, A., Peterseim, D.: Localization of elliptic multiscale problems. Math. Comp. **83**(290), 2583–2603 (2014)

69. Marchenko, V. A. and Khruslov, E. Ya.: Homogenization of Partial Differential Equations. Springer Science & Business Media (2006)

70. Marchenko, V. A., Khruslov, E. Ya.: Boundary-value problems with fine-grained boundary. Mat. Sb. (N.S.) **65**, 458–472 (1964)

71. Maxwell, J. C.: A Treatise on Electricity and Magnetism. Oxford : Clarendon Press (1873)

72. Maz'ya, V., Movchan, A., Nieves, M. and others: Green's Kernels and Meso-scale Approximations in Perforated Domains. Springer, volume 2077 (2013)

73. Michel, J. C., Moulinec, H., Suquet, P.: Effective properties of composite materials with periodic microstructure: a computational approach. Comput. Methods Appl. Mech. Engrg. **172** (1–4), 109–143 (1999)

74. Milton, G. W.: Composite materials with Poisson's Ratios close to −1. J. Mech. Phys. Solids **40**(5), 1105 - 1137 (1992)

75. Milton, G. W.: The Theory of Composites. Cambridge, UK: Cambridge University Press (2002)

76. Müller, S.: Homogenization of nonconvex integral functionals and cellular elastic materials. Arch. Rational Mech. Anal. **99**(3), 189–212 (1987)

77. Murat, F.: Compasite par compensation. Ann. Scuola Norm. Sup. Pisa Cl. Sci. **4**(5), 489–507 (1978)

78. Murat, F., Tartar, L.: H-Convergence. In: Cherkaev A., Kohn R. (eds) Topics in the Mathematical Modelling of Composite Materials. Progress in Nonlinear Differential Equations and Their Applications, vol 31. Birkhäuser, Boston (1997)

79. Murat, F., Tartar, L.: Calculus of variations and homogenization. In: Cherkaev A., Kohn R. (eds) Topics in the Mathematical Modelling of Composite Materials. Progress in Nonlinear Differential Equations and Their Applications, vol 31. Birkhäuser, Boston (1997)

80. Nguetseng, G.: A general convergence result for a functional related to the theory of homogenization. SIAM J. Math. Anal. **20**(3), 608–623 (1989)

81. Nolen, J., Papanicolaou, G., Pironneau, O.: A framework for adaptive multiscale methods for elliptic problems. Multiscale Modeling & Simulation **7**(1), 171–196 (2008)

82. Oleinik, O. A., Shamaev, A. S., Yosifian, G. A.: Mathematical Problems in Elasticity and Homogenization. Elsevier, volume 2 (2009)

83. Owhadi, H.: Bayesian numerical homogenization. Multiscale Model. Simul. **13**(3), 812–828 (2015),

84. Owhadi, H., Zhang, L.: Localized bases for finite-dimensional homogenization approximations with nonseparated scales and high contrast. Multiscale Model. Simul. **9**(4), 1373–1398 (2011)

85. Owhadi, H., Zhang, L., Berlyand, L.: Polyharmonic homogenization, rough polyharmonic splines and sparse super-localization. Published on arXiv:1212.0812 (2012)

86. Owhadi, H., Zhang, L.: Metric-based upscaling. Comm. Pure and Appl. Math. **60**(5), 675–723 (2007)

87. Panasenko, G.: Multi-scale Modelling for Structures and Composites. Springer Netherlands (2005)

88. Pavliotis, G., Stuart, A.: Multiscale Methods: Averaging and Homogenization. Springer (2008)

89. Poisson, S.D.: Memoire sur la theorie du magnetisme. Memoires de l'Academie Royale des Sciences de l'Institute de France **5**, 247–338 (1826)

90. Strutt, J.-W., 3rd Baron Rayleigh 3rd Baron Rayleigh: On the influence of obstacles arranged in a rectangular order upon the properties of a medium. Phil. Mag. **34**, 481–502 (1892)

91. Rockafellar, R.T.: Convex analysis. Princeton university press (2015)

92. Royden, H. L.: Real Analysis. Prentice Hall (1988)

93. Sanchez-Palencia, E.: Non-Homogeneous Media and Vibration Theory (Lecture Notes in Physics). Springer (1980)

94. Shiryaev, A. N.: Probability (volume 95 of Graduate texts in mathematics). Springer-Verlag New York (1996)

95. Sinai, Y. G.: Probability Theory: an Introductory Course. Springer Science & Business Media (2013)

96. Smith, W. A.: Optimizing electromechanical coupling in piezocomposites using polymers with negative Poisson's ratio. In: Ultrasonics Symposium, 1991. Proceedings., IEEE 1991, 661–666 (1991)

97. Tartar, L.: The General Theory of Homogenization: a Personalized Introduction. Springer Science & Business Medi, volume 7 (2009)

98. Torquato, S.: Random Heterogeneous Materials: Microstructure and Macroscopic Properties. Springer Science & Business Media, volume 16 (2013)

99. Volberg, A.: Calderón-Zygmund Capacities and Operators on Nonhomogeneous Spaces. American Mathematical Society (2003)

100. Werner, J.: Potential Theory. Springer-Verlag (1974)

101. Yurinskiĭ, V. V.: Averaging of symmetric diffusion in a random medium. Sibirsk. Mat. Zh. **27**(4), 167–180, 215 (1986)

Index

© Springer Nature Switzerland AG 2018
L. Berlyand, V. Rybalko, *Getting Acquainted with Homogenization and Multiscale*,
Compact Textbooks in Mathematics,
https://doi.org/10.1007/978-3-030-01777-4